Plasticity and failure behavior of solids

Fatigue and Fracture

VOLUME 3

The titles published in this series are listed at the end of this volume.

Plasticity and failure behavior of solids

Memorial volume dedicated to the late
Professor Yuriy Nickolaevich Rabotnov

Edited by

G. C. Sih

Institute of Fracture and Solid Mechanics
Lehigh University, Bethlehem, Pennsylvania 18015, U.S.A.

A. J. Ishlinsky

Institute of Mechanics Problems
USSR Academy of Sciences, Moscow 117526, U.S.S.R.

S. T. Mileiko

Institute of Solid State Physics
USSR Academy of Sciences, Moscow 143432, U.S.S.R.

Kluwer Academic Publishers

Dordrecht / Boston / London

Library of Congress Cataloging-in-Publication Data

CIP

Plasticity and failure behavior of solids . memorial volume dedicated to the late Professor Yuriy Nickolaevich Rabotnov / edited by G.C. Sih, A.J. Ishlinsky, S.T. Mileiko.
p. cm. -- (Series on fatigue and fracture ; v. 3)
Includes bibliographies.
ISBN 0-7923-0336-9 (U.S.)
1. Plasticity. 2. Fracture mechanics. 3. Rabotnov, I͡U. N. (I͡Uriĭ Nikolaevich) I. Rabotnov, I͡U. N. (I͡Uriĭ Nikolaevich) II. Sih, G. C. (George C.) III. Ishlinskiĭ, Aleksandr I͡Ul'evich, 1913- IV. Mileĭko, S. T. (Sergeĭ Tikhonovich) V. Series.
TA418.14.P573 1989
620.1'1233--dc20 89-11175
CIP

ISBN-13: 978-94-010-7332-5 e-ISBN-13: 978-94-009-1866-5
DOI: 10.1007/978-94-009-1866-5

Published by Kluwer Academic Publishers,
P.O. Box 17, 3300 AA Dordrecht, The Netherlands.

Kluwer Academic Publishers incorporates
the publishing programmes of
D. Reidel, Martinus Nijhoff, Dr W. Junk and MTP Press.

Sold and distributed in the U.S.A. and Canada
by Kluwer Academic Publishers,
101 Philip Drive, Norwell, MA 02061, U.S.A.

In all other countries, sold and distributed
by Kluwer Academic Publishers Group,
P.O. Box 322, 3300 AH Dordrecht, The Netherlands.

printed on acid free paper

Fatigue and Fracture

Fatigue and fracture encompass a great many disciplines involving physics, chemistry, continuum mechanics, materials testing and structural analysis. Because of the large number of publications, the analytical and experimental complexities, and the variety of phenomena and materials encountered, it becomes necessary to provide a medium for disseminating pertinent technical material, data, and information on an organized basis.

This new series is devoted to the advancement of theoretical knowledge and practical understanding of fatigue and fracture. It intends to strike a balance between material evaluation and structural design. The ever-increasing demand for high performance structures has necessitated the revision of existing technical principles and the development of new ones. Admittedly, the problem of material failure cannot be completely avoided and will very likely always be with us. Encouraged, however, as contributions to the fundamental understanding and practical application of procedures for design, material selection and fabrication which, when combined into an integrated whole, should provide a better means of evaluating the safety and or reliability of modern engineering structures.

The editorial board

Contents

PREFACE

More than six years ago, several of Rabotnov's close friends and colleagues from the USSR and USA decided to contribute a volume on Plasticity and Failure of Solids in honor of his 70th birthday. The celebration was interrupted unexpectedly by his death on May 13, 1985 at which time another decision was made still to publish the work, but as a memorial volume.

As in any field of scientific endeavor, research confronts the scientists with anomalies; our chosen area is no exception. The ways in which failure criteria and plasticity theory are combined can differ widely among the researchers; they will never yield quite the same results. Each of the invited contributors has, therefore, been encouraged to express his views and to expound on his personal opinion. The contributors are free of enumeration from the authority and/or consensus of any scientific society or community. What impedes scientific process is the esoteric tradition of accepting ideas and theories by consensus among members of societies and communities. The absence of such a trend is refreshing; the collaboration between the authors from the USSR and the USA had to be one of the contributing factors. Finally, the editors wish to acknowledge the authors who have made the publication of this volume possible.

G.C. Sih
S.T. Mileiko
A.J. Ishlinsky

The late Professor Yuriy Nickolaevich Rabotnov
(February 24, 1914 – May 13, 1985)

Scientific biography of the late academician Yu.N. Rabotnov

The late Professor Yuriy Nicholaevich Rabotnov was born on February 24, 1914 in Nigni Novgorod (now Gorky); an old Russian city situated on the Volga River in the very heart of Russia. His father was a teacher and his mother was a medical doctor. He was a member of the USSR Academy of Sciences, an honor bestowed on outstanding USSR scientists in their field of expertise. He passed away on May 13, 1985 after a prolonged illness.

He graduated from the Moscow State University in 1935 and worked as a researcher and lecturer at the Moscow Institute of Energy. His early research and educational works were carried out at the Institute of Mechanics of the USSR Academy of Sciences and at the Moscow State University, then at the Hydrodynamics Institute of the Siberian Branch of the Academy of Sciences. Before his death, Professor Rabotnov held a chair at the Moscow State University and was the head of the fracture mechanics laboratory of Blagonravov's Machinery Institute of the USSR Academy of Sciences.

Scientists of all age groups have high esteem for Rabotnov as a teacher and leader in the field of solid mechanics. His brilliant analytical mind and in-depth intuition on experimentation have provided the basis for many successful scientific achievements. He had the insight to recognize important problems and fulfilled the needs in mechanics. Such a capability has helped to cultivate many outstanding scientists in the USSR.

Rabotnov's research in solid mechanics encompassed a wide range of areas that included plasticity, creep theory, hereditary mechanics, failure mechanics, nonelastic stability, composites and shell theory. It is not possible to describe these activities in detail; only a brief coverage of some of his outstanding work will be mentioned. One of his important contributions is the elastic-plastic hardening models published in 1959 which demonstrated the growth of plastic zones in a thin-walled tube made of an ideal elastic-plastic solid loaded by two bending moments; all the fundamental characteristics of a hardening solid were exhibited. The simplicity of the model revealed the behavior of corner (conical) points on the yield surface and the conditions under which nonincremental plasticity theory remained valid. One must value his foresight in recognizing the effectiveness of such approaches in contrast to some of the direct methods that can be overly complex, particularly, when dealing with composite materials. The paper on 'Equilibrium of an elastic medium with after-effect', which appeared in 1948, is another milestone in mechanics; it renewed the interest in hereditary models of solid state mechanics. This work is well known because of its impact on the development of viscoelastic (hereditary) theories. Two other papers of similar importance were published in 1959 and 1963, dealing with the mechanics of failure of solids. The concept of damage in creep was introduced where the rigidity of a solid could change. Such an idea has been incorporated into the modeling of composites where damage is inherent.

As an educator, Rabotnov lectured widely on many subjects and contributed as an author to a number of books. Among the monographs to be cited are *Creep of Structural Elements* (1966), *Elements of Hereditary Solid Mechanics* (1977) and *Mechanics of Deformable Solids* (1979, 1988). The presentations are laid out systematically and clearly; they can be easily comprehended by a wide range of readers. The first chapter of the book on hereditary mechanics and the introduction to fracture mechanics published in 1979 are worthy of special recognition.

The late Professor Yu.N. Rabotnov will be remembered not only as a distinguished scientist but also as a warm-hearted individual whose sense of humor will live on in the memory of many of his friends around the world. This token contribution by some of his colleagues in the USSR and USA is dedicated to him.

A.J. Ishlinsky
S.T. Mileiko
G.C. Sih

D.C. Drucker

1

Past and possible future directions in plasticity

1.1 Introduction

This is an opportune time in the development of the subject of plasticity, to which Professor Rabotnov has made so many pioneering contributions, for a review of the past, an assessment of the current state of affairs, and a projection into the future. Extensive application now is made routinely of the well-established techniques and concepts of basic plasticity theory to a host of practical problems of analysis and design in aerospace, civil, and mechanical engineering. Engineering practice will continue to absorb the advances made through research as those advances are clarified and sufficiently simplified for use by those who are not experts in the field.

Plasticity theory is a part of the broad and fascinating subject of mechanics of materials or continuum mechanics, which includes elastic and inelastic behavior. The scales of space and time vary enormously from one to another of the important problems. Distances can be as large as those of the giant tectonic plates that move about on the Earth and the length and breadth of the mountain chains that are produced. They drop down to the macroscale of millimeters to kilometers for the small and the large devices, machines, and structures we construct for our use on Earth and in space. Then there is the miniscale on which we tailor materials such as fiber reinforced composites in which the fibers are visible to the unaided eye. Finally we arrive at the microscale or atomic scale of the electron microscope and other devices of enormous magnification.

Research continues to flourish and to branch out in many directions. There is much excitement today in the research community centered on finite elastoplasticity. Also, renewed and expanded efforts are underway to predict the response of structural metals and alloys from dislocation theory and from the calculated properties and the observed behavior of single crystals. Increasing attention is being paid to material behavior at and just prior to failure. Interesting information is becoming available on the behavior of material under extreme conditions of strain rate, temperature, and pressure.

Times of interest range from the eons of geologic time down through years, hours, seconds, microseconds, to picoseconds. Large and obvious time dependency on one scale may be viewed as time independence over another time scale that is very much larger or very much shorter. It certainly is true that time dependence is all-pervasive. Yet, the deformations of a machine or a structure under load at ordinary temperatures can most often be viewed as time independent over ordinary working times. This is so, even though measurable time effects may be obvious during an initial short time of loading, or, at the other extreme, over very long times. Primary, but not exclusive, attention will be devoted here to plasticity in the traditional sense of time-independent behavior of material. It is both proper and useful to analyze and design structures, machines, and devices as time independent under quasistatic loading or at ordinary strain rates in the low to moderate temperature range as is done in engineering practice.

A thread that will run through much of the discussion of plastic stress-strain relations from their origin to the present time is the enormous influence of elastic stress-strain relations. Elasticity has served plasticity alternately to inspire greatly and to mislead terribly as the mathematical structure of elastic stress-strain relations has been refined and generalized. Difficulties arise whenever it is forgotten that, in principle, plastic response bears no resemblance at all in physical or mathematical terms to the most elaborate of possible elastic response or conceptual framework. The essence of plastic behavior is seen most clearly in the limit of vanishingly small – but not ignored – elastic response.

The model of homogeneous plastic response, whether or not there is an accompanying elastic response, is helpful conceptually but is not a direct representation of physical reality. Ordinarily, when plastic deformation takes place in a continuum, most of the volume responds elastically and not plastically. The state of stress and of strain is far from homogeneous in any complex structure or machine or any body under load. When one member or a few members or regions go plastic, most of the other members or regions are below yield. As plastic deformation continues and spreads to other regions, most of the material still responds elastically. At the microstructural level of grains and inclusions, the inhomogeneity of stress and strain is equally marked even when the state of stress and of strain is macroscopically homogeneous. At the atomic or dislocation level, except for the most extreme conditions, only a very small fraction of the atoms move to new positions with respect to their neighbors during a small increment of plastic deformation. Most of the observed plastic deformation at the macroscopic level under an imposed large or small increment of load or displacement therefore has a component of elastic strain on both the microscale and the macroscale that is not recovered upon release of the increment.

Under macroscopically fully plastic conditions, corresponding to limit conditions in the absence of workhardening, the dislocation density changes only moderately, the elastic strains in the continuum also change only moderately, and the macroscopic plastic behavior which is to be modelled is more clearly evident. The plastic deformation that takes place, or is computed to take place, becomes less and less dependent upon the largely unknown initial state and the complex transition from dominantly elastic to dominantly plastic response.

1.2 The beginning of plastic stress-strain relations

The opening paragraph of Professor Rabotnov's paper [1] 'The theory of creep and its applications' sets the stage here with the term 'creep' being replaceable by any inelastic response:

> In choosing a basic system of equations to describe the creep process of machine parts, we must have two objectives in mind. First, the equations must adequately reproduce the experimental data available at the present time and they must at least be free from physical and logical contradictions. Second, the equations must permit the design of rather complicated structures in a sufficiently simple way.

Clarity of physical understanding and the ability to design are primary. It is the essence of the key physical aspects of the relevant behavior that must be captured in an adequate mathematical representation of reality for the purpose in hand.

The basic phenomenological features of plastic behavior, of the theory of plasticity as a useful subject, are: irreversibility, the approximation of time independence, the very low stiffness or resistance to further plastic deformation in ductile metals when compared to the resistance to elastic deformation of the same metal, and finally, the large ductility itself.

The mathematical theory of plasticity, like that of creep which came later, was founded on bold purposeful assumptions. The experiments of Tresca [2] established the concept that large plastic deformation is shear deformation governed by shear stress. It took great insight to ignore the time effects so obvious in some of the experiments, to gloss over the details of the transition from elastic to fully plastic conditions, and to proclaim the governing importance of the maximum shear stress. Coulomb's 'failure' criterion [3], proposed much earlier for soils, appropriately included the strong influence of the normal stress on the shear stress required on the plane sliding.

The plastic stress-strain relations proposed by St.Venant and Levy [4] represented a giant step forward in continuum mechanics. A yield or failure criterion combined with the equations of equilibrium or motion can be used to solve some very interesting 'statically determinate' types of problems [5]. However, without a stress-strain relation, such an approach is both very limited and uncertain in application. The relatively recent development of the theory of elasticity had made the need for stress-strain relations clear. Elastic strains were ignored and plastic strains only were considered. The first principle of fully plastic behavior, beyond the basic phenomenological features, was captured. The ratios of the components of the increments of plastic strain depend primarily upon the stress itself. The increment of stress, which in fact is zero at limit conditions and close to zero at fully plastic conditions, does not even appear in this plastic-rigid model.

$$d\,\varepsilon_x^p = \frac{2}{3}\lambda\left[\sigma_x - \frac{1}{2}(\sigma_y + \sigma_z)\right], \dots , \dots , \tag{1.1}$$

$$d\,\varepsilon_{xy}^p = \frac{1}{2}d\,\gamma_{xy}^p = \lambda\tau_{xy}, \dots , \dots ,$$

or

$$d\,\varepsilon_{ij}^p = \lambda\, s_{ij} = \lambda\left(\sigma_{ij} - \frac{1}{3}\sigma_{kk}\delta_{ij}\right). \tag{1.2}$$

The influence of linear isotropic elasticity is obvious in these expressions with Poisson's ratio set equal to 1/2, but the forms are properly plastic.

Much later, Prandtl and Reuss [6] added an incremental elastic response to the St.Venant–Levy equations to produce the simplest incremental stress–incremental strain relations for an isotropic material. They are mathematically proper and applicable over the entire range, from the elastic response to initial loading all the way through fully plastic conditions, provided the elastic displacements are small and rigid body rotation is absent or properly taken into account.

Meanwhile, and subsequently, many experimenters and those who analyzed their results attempted to refine and understand the yield criterion (or failure criterion as it so often still was called). Maximum energy, maximum shear strain energy, maximum tensile strain, and a host of other criteria were advanced. Mises, while accepting the Tresca criterion or maximum shearing stress as physically correct, proposed that an analytic function of stress be used instead for mathematical convenience. At first he adopted the Huber–Mises–Hencky, or octahedral shear stress, or shear strain energy, or J_2 criterion. Later [7], in analogy with the potential functions which did have a firm physical basis in other branches of applied mathematics and mechanics, he proposed the normality rule for the plastic increment of strain with any choice of a smooth isotropic or anisotropic function which was not necessarily identical to or even similar to the yield function of the material.

1.3 Isotropic hardening as a next step

Once this mathematically and physically permissible, but often overly simple, base had been established for the idealization of stress-strain relations in the plastic range, it became feasible to include more of the clearly significant details of the physical aspects of time-independent material behavior. Workhardening was an obvious candidate with both physical and mathematical questions worth exploring. Again, the major objective was not to encapsulate a vast array of physical knowledge but rather to write as simple a form as possible that would describe the major feature of hardening in mathematical form. A J_2 or Mises isotropic hardening rule [8] did just that for primarily outward loading paths beyond the small strain range. In an approximate way it does reflect the hardening of metallic crystals with increasing immobile dislocation density as shear deformation progresses, a hardening against subsequent *large* plastic deformation for all directions of slip. Of course, it is not realistic to expect agreement in detail between the predictions of so simple a mathematical form and the real behavior of metals or alloys at each point of a general path of loading. The generalization of isotropic stress hardening to include the third invariant of the stress deviation tensor [9] similarly serves a useful but limited role for outward loading paths as does the concept of isotropic expansion of anisotropic yield functions. Wishful thinking would have it otherwise but the complexity of plastic response in detail is far beyond the scope of any explicit mathematical formulation as well as experimental verification.

1.4 Computers and the very limited experimental information obtainable

The details of the response of each metal or alloy to varying load or deformation are extremely complex. A newcomer to the highly developed subject of plasticity often wonders at its apparent backwardness. Why can't all problems be solved directly by writing the constitutive equations in their general form, placing them on the computer, and then simply plugging in the specific properties of the material? That can be done and is done for initially isotropic elastic materials. Why not for plasticity, or at least for initially isotropic elastoplastic material? The answer is that plasticity is not just a more complicated elasticity that requires a more powerful computer and far more experimental data.

Of course, simple idealizations of plastic behavior can be and are placed on the computer and interesting problems are solved with these idealizations. Strong path dependence often requires the taking of very small incremental steps and makes these exercises far from trivial. Even highly idealized problems in two dimensions tax the largest of today's computers. Genuinely three dimensional problems will place great demands on the advanced supercomputers not yet designed, despite drastic idealization far from the reality of material behavior. Furthermore, important aspects of each solution are likely to vary greatly with minor changes in the details of the constitutive equations.

It is not easy to convey the infinite complexity of plastic response. Bridgman spoke of it as a 'sea of irreversibility'. One way of phrasing it is to start with the statement that the only way to determine the incremental stress–incremental strain relation in the plastic region of a deforming body would be to subject a specimen of identical material to the identical loading path. This statement would be followed by a reminder that there is more than enough variability in a most closely controlled material to cause nominally identical material to exhibit small but measurably different response even in simple tension or simple shear. There will be appreciable differences in the individual components of stress and of strain as the same complex loading path is followed for two nominally identical specimens. If results for a number of specimens were to be averaged, the uncertainty in the individual components of stress and of strain calculated for each point in a fully or a partially plastic body under load would be large. The sign of some of the smaller increments of stress or strain might well be in doubt.

Determination of incremental stress–incremental strain relations requires a laboratory specimen that is statically determinate, so that the stresses and the stress increments are determined without ambiguity from the loads and the load increments. The specimen also must be of a geometric shape that permits the determination of strain and strain increments with acceptable accuracy. Therefore, for the discussion here of the experimental determination of stress-strain relations, the most important consideration of all is that there is no possibility of subjecting a test specimen of the material to the general paths of loading to which the material in any small interior region of the body is subjected, none at all. Paths can be followed for regions on the free surface of the body through the use of a thin-walled circular tube specimen, Figure 1.1, but even then only in the unlikely situation of initially isotropic plastic response of the material.

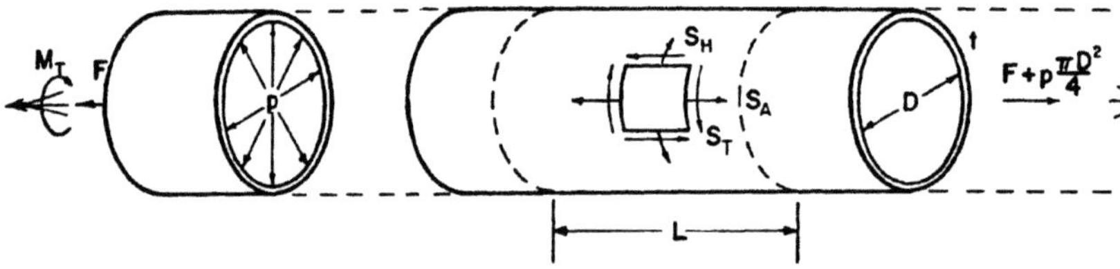

Fig. 1.1: An axisymmetric homogeneous thin-walled tube is statically determinate [30].

As in modern soil and rock mechanics testing it is possible in principle to subject a cube of material to uniformly distributed normal and shear forces on the faces of the cube and follow any stress path on the assumption of homogeneity of stress throughout the specimen. Soil stress levels are low in comparison with the stresses that can be carried by the loading fixtures and rock stresses are moderate so that, with great difficulty, tests can be performed which do apply fairly uniform distributions of surface tractions over much of the area of each of the faces of the cube. Such data is far better than no data at all, but the averaging of inelastic response is a much less suitable procedure than the averaging of elastic response, which in turn is not all that satisfactory in the nonlinear range. The high stress levels of structural metals and alloys in the plastic range makes this type of experimental procedure so difficult that only very crude results could be expected at best.

There is no equivalent of the strain energy density of elastic response which permits a few tests *that can be performed* to determine the full response of an initially isotropic material to any loading. The fact that there is no practicable way at all to perform the full range of tests needed in the plastic regime to determine the complete set of incremental relations between stress and strain is an intrinsic difficulty that cannot be overcome with infinite computer storage and speed of computation.

Consequently it is necessity, not just convenience, that forces idealization of plastic behavior in the absence of experimental information. The salient features must, of course, be captured if the essence of reality is to be portrayed by a mathematical theory of plasticity. These essential features are themselves still a matter of some debate.

1.5 The stability postulate for small displacements

The macroscale and the microscale can be considered at the same time. As indicated schematically, Figure 1.2, stress applied to the microstructure is the analog of load applied to a structure or continuum [10]. If the continuum or structure or microstructure is stable in the strict sense of my definition of requiring positive work by all external agencies that produce small alterations in the deformation or strain of the already loaded body, then all initial and subsequent yield or loading surfaces in load space or in stress space are convex, Figure 1.3, and serve as the analog of potential functions. The corresponding infinitesimal plastic increments of displacement or of plastic strain are normal to the current yield surface. Convexity and normality hold for inhomogeneous as well as homogeneous states of stress and strain. As mentioned earlier, plastic doesn't mean plastic everywhere, but only that the geometry is not fully restored after an incremental loading and release of the added load. The infinitesimal increment of plastic strain or displacement that is normal to the yield surface differs in principle from the increment in permanent or residual strain and will be different in reality if the elastic moduli are altered by plastic deformation [11]

1.6 Perfect plasticity and the limit theorems

Reality is so complex that we must deal with idealizations of idealizations. A single time-independent stress-strain curve in simple tension or in simple shear is an idealization, as is a single load-deflection curve. Yet despite their enormous complexity of response, all ductile metals do exhibit relatively low plastic resistance to increasing load. This is the feature captured in exaggerated but very useful form by the further idealization of ideal or perfect plasticity. For problems of cold metal deformation processing, with very large plastic strains, the flow strength can be chosen appropriately high. For the small and moderate strain range, for most engineering designs, the flow strength of the very same metal or alloy should be chosen to be quite a bit smaller. It is worth emphasizing over and over again that the idealization selected depends fully as much on the problem to be solved as on the material itself.

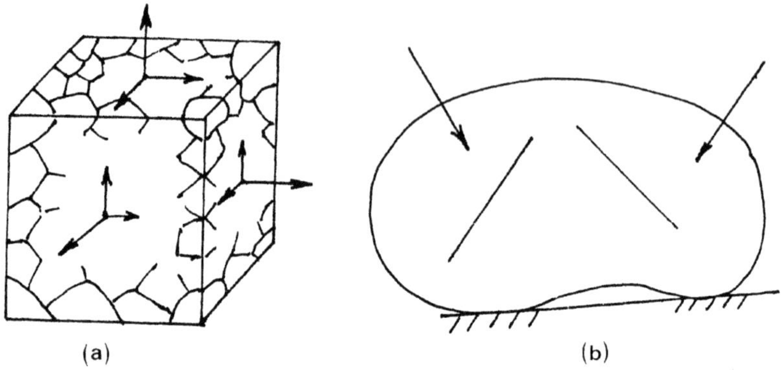

Fig. 1.2: Microstructure and structure or continuum [10]. (a) macrostress on microstructure; (b) loads on structure or continuum.

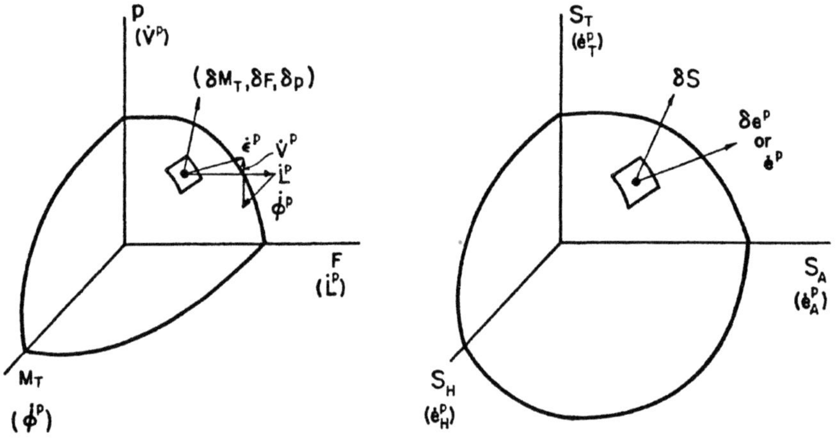

Fig. 1.3: Normality and convexity [30].

Convexity, normality, and the idealization of perfect plasticity give the limit theorems of plasticity. They are theorems which agreeably match the engineer's intuition and insight; theorems with comforting consequences, such as the absence of any influence of initial or residual stress on the limit or failure load. They do hold for plastic material but unfortunately not for frictional material [12], although friction is sometimes viewed as the analog of plasticity. So much for wishful thinking. Nature is often perverse. Initial stresses, tectonic stresses for example, can have devastating effects.

The limit theorems provide an excellent first order approximation for all ductile behavior on the macroscale and the microscale when the reality of workhardening is kept in mind along with the reality of dislocations. Many useful results, many ways of thinking about machines, structures, and materials as continua have come out of the plastic limit theorems [10, 13].

1.7 Stress invariants and further remarks on idealization and reality

Isotropy of elastic response is very close to reality for most polycrystalline metals with randomly oriented crystals and so provides a very suitable start for elastic stress-strain relations. Anisotropy of incremental elastic response will occur in a highly stressed isotropic material due to the existing state of stress, but the total response is independent of direction in the material at all stress levels for which the material is stable. Elastic response is path independent so that isotropic elastic material remains isotropic. Initial isotropy of plastic response on the contrary is very uncommon and can be achieved only through the greatest of care in the preparation of the material and in the fabrication of the test specimen. Furthermore, just about any subsequent plastic deformation destroys the isotropy. The suitability of isotropy and the use of stress invariants as a point of departure for stress-strain relations in the plastic range lies in the resulting mathematical simplicity, not in physical reality.

Osgood, some 40 years ago, took the trouble to obtain and test an initially isotropic structural aluminum alloy in the form of a thin-walled tube under combinations of interior pressure and axial tension [14]. He plotted his test results as maximum shear stress vs. shear strain and octahedral shear stress (or the second invariant of the stress deviation tensor) vs. the corresponding shear strain. Both plots looked very good with an apparent scatter under 10%. If Osgood were not so careful an experimenter there would have been no point to any further analysis. It took a particular combination of the second and third invariants of the stress deviation tensor to show that the data in the plastic range could indeed be placed on a single curve [15]. That success did not indicate that in fact those two stress invariants controlled plastic deformation, but simply confirmed that shear stress in a generalized sense controlled plastic deformation. It is a purely mathematical requirement [16] that the plastic response to radial or proportional loading (all stress components increasing in fixed ratio) be correlated by those two invariants for an initially isotropic material insensitive to hydrostatic or all-around pressure, when the ratios of the plastic strain increments do not change appreciably along a radial path. The correlation in no sense can be taken to mean that the response to nonradial or general loading paths is similarly governed. The so-called isotropic hardening plasticity forms based on invariants do not depict physical reality at all closely except for paths of loading that are not too far from radial.

We know that information obtained from radial loading tests, three independent radial loading tests, does determine the response of a nonlinear elastic material to any path of loading whatsoever, no matter how wiggly or nonradial. It is understandable, therefore, but always regrettable in retrospect, that the vain search for equal simplicity of plastic response keeps re-emerging in one form or another. Drastic idealization, repeated frequently enough, all too often becomes confused with reality.

The unjustified hope that plasticity really can't be that much more complicated than elasticity led many years ago to a decade of fighting over the validity of so-called deformation or total theories of plasticity. These are simply nonlinear elasticity forms which match the stress-strain curves for simple tension or simple shear or any radial loading without unloading. There can be no quarrel with their use for convenience for any generally outward loading path which does not deviate too much from radial. In fact they are in common use for elastoplastic fracture mechanics because they are simple enough to permit the solution of problems and they should provide more suitable indications of the surround of a crack tip than the elastic solutions of linear elastic fracture mechanics. However, no one argues any longer that, in principle, deformation theory is a proper representation of plastic behavior for all paths of continued loading.

1.8 Finite elastoplasticity for small incremental displacements

Today we argue about finite elastoplasticity instead. The discussion is on a far more sophisticated mathematical and physical level. Yet a noisy debate will probably go on for several more years. It began in 1981 at a workshop convened by Professor E.H. Lee to resolve the issue of the definition of plastic strain for large displacements. The focus of attention shifted to the surprising oscillating shear stress results of computer calculations for very large shear strains reported by Drs. Nagtegaal and de Jong [17], Figure 1.4. They used a computer program, typical of most computer programs then – typical of almost all computer programs now – that dealt with finite deformations. These programs have embedded in them concepts thought suitable for elastic response, concepts which are fundamentally not suitable for plastic response.

Discussions, meetings, and papers in the years following the workshop have clarified a number of the issues. In the first place, a similar unrealistic oscillation of the shear stress with increasing shear strain is obtained from these same computer programs for what is supposed to be purely elastic response. The oscillation is an artifact of writing incremental stress-incremental strain relations for the Cauchy stress components referred to axes rotating without limit at the Jaumann rate of spin. It occurs as the reference axes are repeatedly rotated through 360 degrees. Such incremental stress-strain relations are not truly elastic in that they do not derive from a strain energy density. They are termed hypoelastic. They are convenient because they are simple. They are appropriate for infinitesimal displacements and can be a very good approximation for moderate finite strains and rotations.

For a nonlinear elastic material, there is no difficulty in principle, and no overwhelming difficulty in practice, in foregoing the use of a simple hypoelastic form and using a proper elastic stress-strain relation instead. The rotation, instead of being the sum of the Jaumann spins, which is unbounded, becomes the familiar continuum rotation for finite displacement. When that is done, the stress-strain curve takes a correct and proper non oscillatory shape.

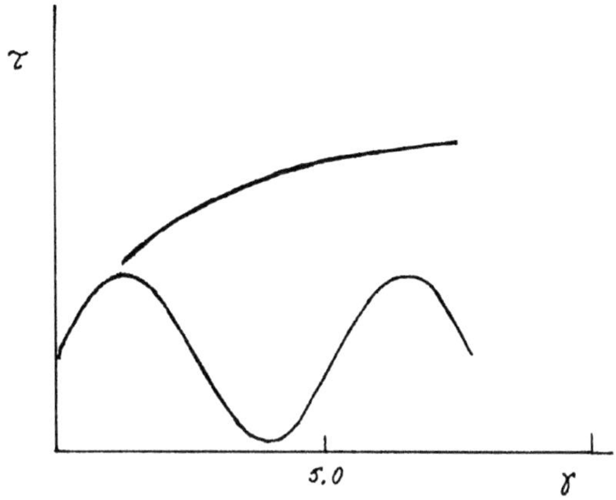

Fig. 1.4: A problem with computer results [17, 18]. Upper curve for isotropic hardening, lower for kinematic.

When a similar approach is taken for an elastoplastic material, the Jaumann spin replaced by the increment in the continuum rotation, and the incremental stress-strain relation assumed to hold in that rotated configuration, a reasonable looking rising stress-strain curve also is obtained [18]. The unpleasant oscillation is gone. In my view, however, all is not well. The continuum rotation due to shear deformation is real and for isotropic elastic material is a physically meaningful rotation that must be followed if the intrinsic simplicity of elastic response is to be exhibited. The continuum rotation, due to plastic shear deformation, is real but irrelevant. There is neither an experimental basis nor a compelling physical reason for writing incremental stress – incremental plastic strain relations for increments in Cauchy stress rotating with the continuum rotation. Once again elasticity misleads when followed too closely. The only experimental information available is for conventional stress in a fixed coordinate system; the axial, circumferential, and radial directions in a circular bar or a thin-walled tube, Figure 1.1. The data could, of course, be converted to rotating axes and then converted back again, but that would serve no real purpose. If there is an instability, and it is the prospect of such instability or localization that accounts for the great interest in the topic, that instability would show up in the original data. An instability found by assuming the data obtained for fixed directions to apply in rotating axes in the same specimen or situation is not to be taken seriously. However, any rigid body rotation, as in the torsional buckling of thin-walled sections or plates, that alters the stresses produced by the applied loads, must be taken into account [19]. So also must the rotation along the flow lines in metal deformation processing be followed for a material that is not assumed to harden isotropically. For the specimen of Figure 1.1, and for most continuum boundary value problems, these rotations are zero when the elastic strains are very small. No matter how large the plastic shearing may be in a circular tube or bar, the physical rotation of the material with respect to the circumferential direction is zero to a first approximation. The lattice planes will on average exhibit this zero rotation [20].

Yet, because of the intensive and important debate about likely instabilities or localization, the response of highly deformed ductile materials to subsequent increments of plastic deformation does warrant some additional study. In particular, the response of heavily sheared material, as in a highly twisted torsion specimen, to combinations of axial stress and shear stress, could be obtained by twisting a solid circular bar to very large shear strain under a small stabilizing axial tension, and then machining a concentric thin-walled tube from it. The incremental plastic response to incremental loading would provide the data needed on stability or instability. However, it would not settle the argument about whether it is 'better' to deal with increments of stress and of strain in a coordinate system that rotates with the continuum rotation of the material or to deal with the conventional increments of stress and strain in the fixed reference configuration, the nonrotated system.

Even if the material is found to be stable under a combined axial compression and torsional shear load, and the effective shear modulus for incremental shear stress–incremental shear strain on the reference planes is positive for conventional fixed axes and negative for the rotated system, nothing really is settled. At a given twisting moment, decreasing conventional shear stress with increasing shear strain does represent instability. Decreasing shear stress in a rotating system, when other components of stress are present, need not be an instability.

The argument, if there is an argument still, is on the matter of convenience of representation, not necessity. One set of incremental stress–incremental strain relations is uniquely translatable to the other. The type of question that might be resolved by experiment is whether, after finite increments of twist, simple kinematic hardening [21, 22] in the conventional system of axes or in the rotated system provides a better physical description of the behavior of the material. The lack of relevance of the continuum rotation associated with the finite shear deformation is obvious for the isotropic hardening idealization. The response to an increment of shear stress after finite prestrain is no different for prestrain in tension or prestrain in shear at the same level of hardening. For the kinematic hardening idealization, the basic question is whether the 'back stress', after finite shear produced by pure twisting moment in the absence of axial force, is more closely related to the rotated directions of principal strain or to the fixed circumferential direction.

Unfortunately, it is all too likely that the real behavior of any material is so far from simple kinematic hardening, and the current yield surface so much a function of the offset plastic strain by which it is defined, that the data will not be as helpful as desired. Perhaps the most to be expected is that a direct test can be made at the stress and deformation states present prior to observed localization in metal deformation processing to determine whether the instability is due primarily to adiabatic heating or is basically a mechanical instability that occurs also under isothermal conditions.

There is much more to be said on the topic and many people are saying it. My prediction is that when the dust settles, the consensus for metals will be to ignore the continuum rotation associated with plastic shearing and continue to use the ordinary incremental stress-strain relations of plasticity, as in the past, taking genuine rigid body rotations of the material into account. We should not forget that the initial formalization by St.Venant and Levy was inspired by the experiments of Tresca in which the shear deformations were very large.

1.9 On some future directions and other concluding comments

Many topics are well worth pursuing and all researchers have their own lists. Mine includes far more than I can list here.

Should it become clear that there is a class of materials for which the combination of large elastic shear distortion with continuing plastic deformation is a physically meaningful and important problem, finite elastoplasticity will then warrant considerable research attention. Experiments to establish aspects of real behavior will be needed to provide benchmarks against which predictions of mathematical idealizations can be compared. Polymeric materials may lie in this class, but their time-dependent behavior is likely to confuse the data and the interpretations of the experiments. Ultrahigh-strength materials, on the other hand, tend to have little macroscopic ductility in the absence of very high hydrostatic pressure.

Non-associated flow rules, that are connected with pressure dependence of the flow stress in shear, call for much further experimental and theoretical study. So also do associated as well as non-associated flow rules based upon thermodynamic assumptions such as maximum entropy production [23].

Cyclic loadings combined with other loadings can now be explored fairly easily with modern computer controlled testing machines. More realistic idealizations than the one proposed with Luc Palgen [24], and those proposed by others, should now be formulated to deal adequately with complicated loading paths and yet still be tractable for practical use.

A much broader topic, on which considerable progress has been made in recent years, following the generalization by Professor Rabotnov [25] of a scalar measure proposed by Professor Kachanov [26], is the study of and the mathematical formulation of the properties of materials in the final stages of failure. Design against failure is rather incomplete if it stops short of considering the safety of the response of these final stages [27]. Dynamic overloads as well as static loads that can be carried in the failed state are of considerable interest.

A broader topic still is the design of materials themselves on the microscale to achieve the macroscopic mechanical properties desired (yield and flow strengths, fracture toughness, ductility, etc.). It is generally true, even for fiber reinforced composites, that the engineer is at best given a choice of materials designed or invented by others. How much better it would be if enough were known to design the materials to match the requirements at hand, and of course to produce the material at affordable cost. Great progress has been made in the application of mechanics on the microscale in a number of instances, as we have seen. Yet in terms of the complexity of materials, no more than a bare beginning can be claimed. There is much to be learned and much to try out before significant advances can be made over the present, primarily trial and error but rather successful approaches, based upon the principles of metallurgical, ceramic, and polymer science.

The microscale is most fascinating. Very simple calculations of average shear stress, shear force divided by area, have provided an understanding of the scales of importance and the possible effects of inclusions and precipitates. Here is where dislocations and dislocation theory calculations interact with continuum calculations to the benefit of both if an open mind is kept on each of the issues that may be controversial at the time.

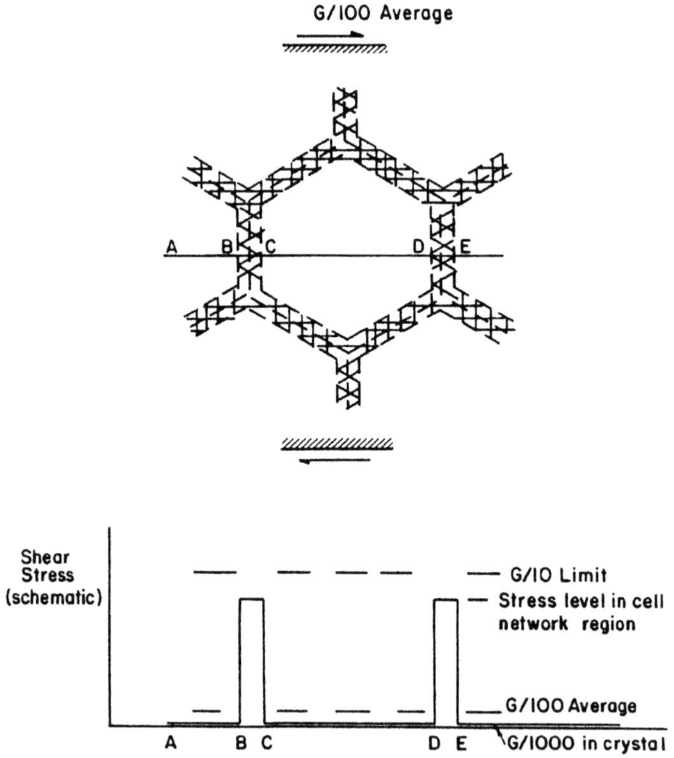

Fig. 1.5: Schematic representation of dislocation cells and of stress levels [28].

Figure 1.5 exhibits some of the scope of the needed understanding of the macroscopic response of metals and alloys in terms of the microscopic [28]. The shear stress level for easy glide in a face centered cubic crystal like aluminum is less than $G/10000$, where G is the elastic shear modulus. As indicated schematically, after a rather dense pattern of intersecting slip has been established in a single crystal, the stress level tops out at about $G/1000$. A structural aluminum alloy has a flow strength in shear of almost $G/100$ at low to moderate strain rates. This value rises sharply in the region of ultrahigh strain rates. The limit of atomic lattice stability occurs at a stress level of about $G/10$ with a nonlinear elastic strain of about 20%. All of these numbers are rather crude but the 1000 : 1 ratio from easy glide to structural alloy response is not an exaggeration.

Suppose the schematic picture conveys the essence of physical reality, whether the dislocation lines are spread more uniformly through the bulk of the material or are in the tangled cell network seen in the electron microscope in the absence of bulk constraint. How then do the basic calculations for stress levels, strains, and rotations in single crystals, bicrystals, and simple polycrystals of aluminum and similar ductile metals fit in? The calculations do predict the results of simple experiments on such crystals and are a great aid in understanding more complex experiments [29]. Yet they would appear to be applicable only in the regions of relatively low dislocation density and low maximum shear stress. Their relation to the key aspects of the behavior of structural metals at 10 times the level of maximum shear stress is far from obvious. Many exciting areas of research still lie ahead combining the viewpoints of continuum mechanics, dislocation mechanics, and the mechanics of disordered atomic arrays.

There is so much to think about, so much to do, that a busy, productive, and exciting future is in prospect for all who wish to participate in this and the many other fascinating areas of research in both the time-independent and time-dependent domain that Professor Rabotnov and others have described elsewhere.

Note added in proof

Some of the material presented here has been printed elsewhere in the time that has elapsed. In particular, significant portions appear verbatim in my Applied Mechanics Reviews' article "Conventional and Unconventional Plastic Response and Representation", Vol. 41, 1988, pp. 151–167.

References

[1] RABOTNOV, Y.N., The theory of creep and its applications, *Plasticity,* eds. E.H. Lee and P.S. Symonds, Pergamon Press, 1960, pp. 338–346.

[2] TRESCA, H., Mémoire sur le poinconnage des métaux et des matières plastiques, *Comptes Rendus,* Paris, Vol. 70, 1870, pp. 27–31.

[3] COULOMB, C.A., Essai sur une application des règles de maximis et minimis à quelque problèmes de statique, relatifs à l'architecture, *Mémoires de Mathématique de l'Académie Royale des Sciences,* Paris, Vol. 7, 1773, pp. 343–382.

[4] LÉVY, M., Mémoire sur les équations générales des mouvements intérieurs des corps solides ductiles au delà des limites où l'élasticité pourrait les ramener à leur premier état, *Comptes*

Rendus, Paris, Vol. 70, 1870, pp. 1323 – 1325.

[5] SOKOLOVSKY, V.V., *Statics of Soil Media,* Butterworth, 1960; *Theory of Plasticity,* Moscow, 1946.

[6] REUSS, A., Berucksichtigung der elastischen Formänderung in der Plastizitätstheorie, *ZAMM,* Vol. 10, 1930, pp. 266 – 274.

[7] MISES, R. VON, Mechanik der plastischen Formänderung von Kristallen, *ZAMM,* Vol. 8, 1928, pp. 161 – 185.

[8] ODQVIST, F.K.G., Die Verfestigung von flusseisenähnlichen Körpern. Ein Beitrag zur Plastizitätstheorie, *ZAMM,* Vol. 13, 1933, pp. 360 – 363.

[9] MELAN, E., Zur Plastizität des räumlichen Kontinuums, *Ingenieur-Archiv,* Vol. 9, 1938, pp. 116 – 126.

[10] DRUCKER, D.C., The continuum theory of plasticity on the macroscale and microscale, *Journal of Materials ASTM,* Vol. 1, 1966, pp. 873 – 910.

[11] PALMER, A.C., MEIER, G., and DRUCKER, D.C., Normality relations and convexity of yield surfaces for unstable materials or structural elements, *Journal of Applied Mechanics ASME,* Vol. 34, 1967, pp. 464 – 470.

[12] DRUCKER, D.C., Coulomb friction, plasticity, and limit loads, *Journal of Applied Mechanics ASME,* Vol. 21, 1954, pp. 71 – 74.

[13] BUTLER, T. and DRUCKER, D.C., Yield strength and microstructural scale: a continuum study of pearlitic vs. spheroidized steel, *Journal of Applied Mechanics ASME,* Vol. 40, 1973, pp. 780 – 784.

[14] OSGOOD, W.R., Combined-stress tests on 24S-T aluminum alloy tubes, *Journal of Applied Mechanics ASME,* Vol. 14, 1947, pp. A147 – A153.

[15] DRUCKER, D.C., Relation of experiments to mathematical theories of plasticity, *Journal of Applied Mechanics ASME,* Vol. 16, 1949, pp. A349 – A357.

[16] PRAGER, W., The stress-strain laws of the mathematical theory of plasticity. A survey of recent progress, *Journal of Applied Mechanics ASME,* Vol. 15, 1948, pp. 226 – 233.

[17] NAGTEGAAL, J.C. and DE JONG, J.E., Some aspects of non-isotropic workhardening in finite strain plasticity, *Plasticity of Metals at Finite Strain: Theory, Computation and Experiment* eds. Lee, E.H., and Mallett, R.L., Stanford University and Rensselaer Polytechnic Institute, 1982, pp. 65 – 101.

[18] LEE, E.H., MALLETT, R.L. and WERTHEIMER, T.B., Stress analysis for anisostropic hardening in finite-deformation plasticity, *Journal of Applied Mechanics ASME,* Vol. 50, 1983, pp. 554 – 560.

[19] ONAT, E.T. and DRUCKER, D.C., Inelastic instability and incremental theories of plasticity, *Journal of the Aeronautical Sciences,* Vol. 20, 1953, pp. 181 – 186.

[20] DRUCKER, D.C., Appropriate simple idealizations for finite plasticity, *Plasticity Today: Modelling, Methods, and Applications,* eds. Sawczuk, A. and Bianchi, G., Elsevier, 1985, pp. 47 – 59.

[21] PRAGER, W., The theory of plasticity – a survey of recent achievements, *Proc. Inst. Mechanical Engineers,* Vol. 169, 1955, pp. 41 – 57.

[22] SHIELD, R.T. and ZIEGLER, H., On Prager's hardening rule, *ZAMP,* Vol. 9a, 1958, pp. 260 – 276.

[23] ZIEGLER, H., *An Introduction to Thermomechanics, 2nd Ed.* North Holland, 1983.

[24] PALGEN, L. and DRUCKER, D.C., On stress-strain relations suitable for cyclic and other loading, *Journal of Applied Mechanics ASME,* Vol. 48, 1981, pp. 479–485.

[25] RABOTNOV, Y.N., *Creep Problems in Structural Members,* North Holland Press, 1969.

[26] KACHANOV, L.M., Time of the rupture process under creep conditions, *Izv. Akad. Nauk. USSR, Otd. Tekh, Nauk,* Vol. 8, 1958, pp. 26–31.

[27] DRUCKER, D.C., Some classes of inelastic materials-related problems basic to future technologies, *Nuclear Engineering and Design,* Vol. 57, 1980, pp. 309–322.

[28] DRUCKER, D.C., Material response and continuum relations: or from microscales to macroscales, *Journal of Engineering Materials and Technology, Trans. ASME,* Vol. 106, 1984, pp. 286–289.

[29] ASARO, R.J., Micromechanics of crystals and polycrystals, *Advances in Applied Mechanics,* Vol. 23, Academic Press, 1983, pp. 1–115.

[30] DRUCKER,D.C., From limited experimental information to appropriately idealized stress-strain relations, *Mechanics of Engineering Materials,* eds. Desai, C.S. and Gallagher, R.H., John Wiley & Sons Ltd., 1984, pp. 231–251.

T.H. Lin and G.E. Ribeiro

2

Physical theory of plasticity: a multicrystal model

2.1 Introduction

Single crystal tests [1, 2] have shown that under stress, slip occurs along certain crystal directions on certain crystal planes and slip depends on the resolved shear stress and is independent of the normal stress on the sliding plane. This resolved shear stress that initiates or causes the continuation of slip is called the critical shear stress, which varies with the amount of slip. After the relation between the resolved shear stress and slip was experimentally determined for single crystals, many early attempts were made to deduce the uniaxial stress-strain relation of a polycrystal from the stress-strain relation of single crystals. The first realistic model was proposed by Taylor [2]. He assumed all grains to have the same homogeneous strain as that imposed on the aggregate and also assumed the crystals to be rigid-plastic. The neglect to elastic strain in Taylor's model gives significant error when the elastic strain is of a comparable magnitude to the plastic strain. Lin [3] extended Taylor's model to include elastic strain. Recently, Tokuda, Kratochvril and Ohashi [4] also assumed uniform strain, elastic plus plastic in a simplified two-dimensional polycrystal and calculated the variation of macroscopic stress under some arbitrary strain paths. The calculated values were found to agree well with experimental results. All these theories satisfy the condition of compatibility but not the condition of equilibrium across the grain boundaries.

Eshelby in 1957 [5] has shown in his interesting paper that an ellipsoidal inclusion in an infinite homogeneous elastic medium to undergo a change in shape and size that would be an arbitrary homogeneous strain if the surrounding material were absent, will cause a uniform strain inside the inclusion. Kroner [6], based on this result, proposed an important approach to the calculation of the stress-strain relation of polycrystals. He considered each crystal in a uniformly loaded polycrystal to be an inclusion in a homogeneous infinite elastic medium. Uniform slip occurs in the crystal when the resolved shear stress exceeds the critical shear stress. The resultant of the loads carried by all the individual crystals cut by a section must balance the applied load on the aggregate. The stress relieved by the slipped crystals must be absorbed by other crystals. Therefore, the sliding of one group of crystals increase the average load taken by other groups of crystals. Taking this average interaction effect between groups of crystals into consideration, Kroner developed an analytical procedure to calculate the polycrystal stress-strain relationship from single crystal characteristics. Budiansky and Wu [7] rederived Kroner's scheme using a different physical reasoning. This scheme is well known as the self-consistent method. When slip has occurred in a significant portion of the aggregate, the matrix of the inclusion has pronounced directional weakness. A theory considering this directional weakness has been given by

Hershey [8] and Hill [9]. Hutchinson [10] made significant contributions to this theory. He calculated the incremental stress-strain relations at different ratios of incremental shear and tensile stresses after being stressed in tension beyond the elastic range. Recently, Weng [11] has applied this self-consistent method to the study of the creep behavior of metals and has made significant contributions.

In the self-consistent method, each crystal is considered as an inclusion in the calculations of the amount of slip in the crystal. It is also considered as a part of this matrix when slip is calculated for any other crystal. Hence, the stress in a crystal should be the sum of stress as an inclusion and as part of the matrix. But the stress in the role of matrix is not explicitly considered to cause slip [10]. The average interaction action of slipped crystals provided by the self-consistent approach does not consider the relative positions of the slipped crystals. Slip varies nonlinearly with the resolved shear stress. This nonlinear effect introduces errors in using this average interaction effect in the calculations of plastic strains.

In most numerical calculation in this approach, spherical inclusions were considered. The resolved shear stress $\Delta\tau$ relieved in such an inclusion caused by a uniform plastic shear strain $\varepsilon''_{\alpha\beta}$ is $2\mu(1-b)e''_{\alpha\beta}$ where $b = 2(4-5\nu)/15(1-\nu)$, ν is Poisson's ratio and is generally taken as 0.3, μ is the shear modulus. This gives an $\Delta\tau = 0.524 \times 2\mu\varepsilon''_{\alpha\beta}$. The plastic strain distribution in an inclusion of cubic shape to relieve a constant resolved shear stress $\Delta\tau$ was shown by Lin *et al.* [12] and is given in Figure 2.1. The average plastic strain to relieve 0.1 C_0 stress is about 0.14 $C_0/2\mu$. This gives $\Delta\tau = 0.71 \times 2\mu\varepsilon''_{\alpha\beta}$. It is seen that spherical inclusions are much softer than cubic inclusions. From the microscopic pictures of metals, the crystals are of polygonal shape. Besides, three dimensional space cannot be filled by ellipsoids alone. This assumption of ellipsoidal crystals is not quite realistic. Hence, a more rigorous theory is desirable.

2.2 Requirements of multicrystal model

The physical theory developed by Lin *et al.* [13, 14] is much more rigorous. The relative positions of the slid crystals are fully considered. The aggregate is considered to be composed of a number of crystals of cubic shape. These clearly fill the three dimensional space well.

Metals are generally composed of a very large number of crystals of random orientations. Due to this large number of differently oriented crystals, the polycrystals have the following general deformation characteristics.

(1) Under uniaxial loading S_{11}, the plastic strain components are found to have the following relation:

$$E''_{22} = E''_{33} = -\frac{1}{2}E''_{11} \tag{2.1}$$

$$E''_{12} = E''_{23} = E''_{31} = 0. \tag{2.2}$$

Similar relations hold for axial loading S_{22} and S_{33}. The relations E''_{11} vs. S_{11}, E''_{22} vs. S_{22} and E''_{33} vs. S_{33} are the same.

(2) Under uniaxial shear loading S_{12}, only the corresponding plastic shear strain E''_{12} occurs. Other plastic strain components vanish. The relations E''_{12} vs. S_{12}, E''_{23} vs. S_{23} and E''_{31} vs. S_{31} are the same.

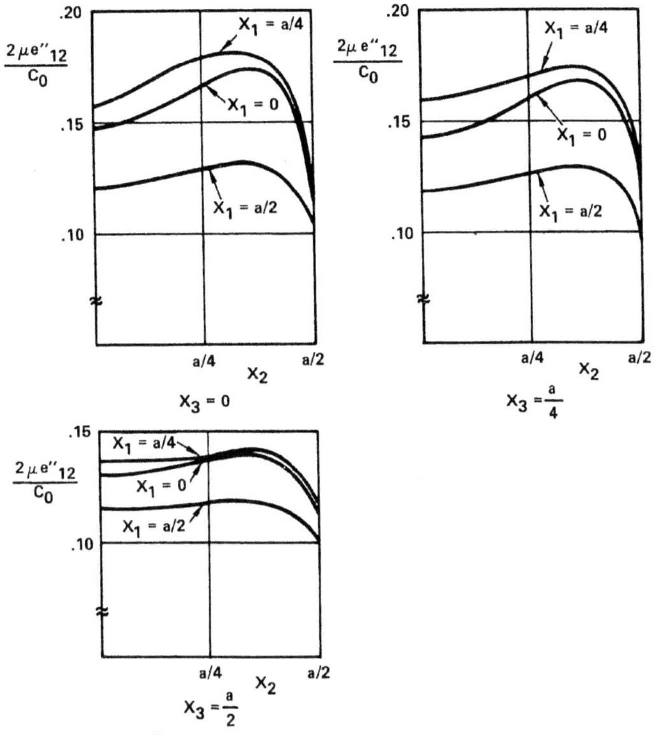

Fig. 2.1: Distribution of slips e''_{12}. Note: The ratio of the resolved shear stress T_0 to the critical shear stress C_0 is 1.10.

These relations are here called the initial isotropy. The self-consistent method has the advantage of satisfying these conditions automatically. In general, the reference axes of this stress and strain can be arbitrarily rotated without affecting the above relations. In our development of the plasticity theory, we consider these reference axes to be fixed. The above relations then correspond to cubic symmetry [12].

2.3 Multicrystal model

Consider a cubic block of 64 crystals as shown in Figure 2.2 Those crystals are divided into 8 groups of 8 crystals each. The first group occupies the first octant. The second group is oriented as an image of the first with respect to a mirror in the x_1x_3 –plane. The 3rd and 4th groups are oriented as the image of the first two groups with respect to an image of a mirror in the x_2x_3 –plane. Finally, the remaining four groups are then oriented as the image of the first four groups with respect to a mirror in the x_1x_2 –

plane. Thus, we have three planes of symmetry and the conditions of orthotropy is thus obtained.

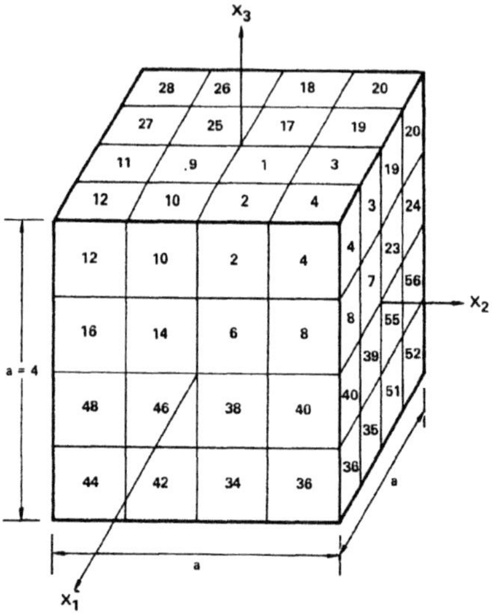

Fig. 2.2: Basic block of 64-cubic-shaped crystals and the crystal numbers.

Now we need only to determine the orientations of the 8 crystals of the first group. Let α denote the normal to the slip plane and β the slip direction of a slip system. Let x_i denote the specimen axis. The direction cosines of α and β are denoted by $l_{\alpha i}$, $l_{\beta j}$ with respect to the specimen axes. Let crystal 'K' be the most favorably oriented crystal under S_{11} and S_{12}. Let the mth slip system be the most favorably oriented under S_{11} and the nth slip system be that under S_{12}. For the mth system to slide under S_{11} we have $(l_{\alpha 1} l_{\beta 1})_m S_{11} = \tau_c$. For the nth system to slide under S_{12}, we have $(l_{\alpha 1} l_{\beta 2} + l_{\alpha 2} l_{\beta 1})_n S_{12} = \tau_c$. We let this crystal be oriented to give $(l_{\alpha 1} l_{\beta 1})_m = 0.577(l_{\alpha 1} l_{\beta 2} + l_{\beta 1})_n$. This gives an initial yielding with a yielding shear stress S_{12} equal to $0.577 \times$ yielding stress S_{11} in tension. This corresponds to Von Mises' criterion of initial yielding. If we change the value 0.577 to 0.500, we obtain Tresca's criterion of yielding. If polycrystal tests show an initial yielding between the Von Mises and Tresca values, this crystal can also be oriented to simulate the case.

Now consider the case under S_{11}, the mk slip system of this crystal slides to give a plastic strain γ. The microscopic plastic strain caused by this slip is then

$$E''_{11} = \frac{1}{64}(2l_{\alpha i} l_{\beta i})\gamma_{mk} \quad E''_{22} = \frac{L}{64}(2l_{\alpha 2} l_{\beta 2})\gamma_{mk}$$

$$E''_{33} = \frac{1}{64}(2l_{\alpha 3} l_{\beta 3})\gamma_{mk}. \tag{2.3}$$

The above satisfies the condition of zero plastic dilatation. Now we let another crystal p be oriented to have

$$\begin{aligned}(l_{\alpha 1} l_{\beta 1})_{np} &= (l_{\alpha 1} l_{\beta 1})_{mk} \\ (l_{\alpha 2} l_{\beta 2})_{np} &= (l_{\alpha 3} l_{\beta 3})_{mk} \\ (l_{\alpha 3} l_{\beta 3})_{np} &= (l_{\alpha 2} l_{\beta 2})_{mk}.\end{aligned} \tag{2.4}$$

Physically, this means that crystal p is the image of crystal k with respect to a mirror in a plane containing the x_1 –axis and forming an angle of 45° with x_2 and x_3 axes. These two crystals are placed in the first octant symmetrical with respect to x_2 and x_3 axes, these two crystals are marked as #3 and #6 in Figure 2.2. Under S_{11}, $\gamma_{mk} = \gamma_{np}$. Equation (2.1) is then satisfied. Due to orthotropy Equation (2.2) is also satisfied. Similarly, corresponding to S_{22}, we have two crystals #2 and #7; correspondingly for S_{33}; we have crystal #1 and #8. Crystals #4 and #5 are oriented to have the crystal axes coinciding with the specimen axis. In this way the condition of cubic symmetry is completely satisfied. The orientation of the first 8 crystals are given in Table 2.1. Before this 64-crystal model was developed, we used a 27-crystal model with orientations of these 27 crystals fairly uniformly distributed over a stereographic triangle. It was found that it is very difficult to fulfill the conditions given by Equations (2.1) and (2.2). In order to use this physical theory to predict polycrystal deformation characteristics, it is essential to satisfy these conditions. Hence, this model is essential for practical applications such as structural analysis.

2.4 Some numerical results using this model

Now if this block of 64-crystals is loaded under a uniform macroscopic loading, heterogeneous slip will occur. To calculate the stress and strain fields in the block, we consider this heterogeneous slip as applied forces [15]. This would given an equivalent problem of a three dimensional elastic solid subject to given forces. This is a three dimensional boundary problem, which is known to be a big task. In order to avoid this task, we consider an infinite number of these basic cubic blocks embedded in an elastic infinite medium with the same elastic constants. This infinite number of cubic blocks can be considered as an inclusion (Figure 2.3) and the forces can be considered to act in an infinite elastic medium.

The solution of the stress field for this medium was given by Kelvin a long time ago. The average stress and plastic strain in the center blocks are here taken to be the macroscopic stress and strain. The use of an infinite number of identical cubic blocks instead of just one single block is to make the surrounding regions of the center block more flexible so as to accommodate more plastic strain.

Using the analogy of plastic strain gradient and applied force and Kelvin's solution, Lin and Ribeiro [16] have shown that the residual stress field caused by a plastic strain field can be written as

$$\tau^{R}_{ijqn}(p) = a_{ijqn}(p)\varepsilon^{p}(q, n) \tag{2.5}$$

where $\varepsilon^{p}(q, n)$ is the plastic strain caused by slip in the nth slip system of the qth crystal. $a_{ijqn}(p)$ is the stress τ_{ij} in the pth crystal caused by a unit plastic shear strain $\varepsilon^{p}(q, n)$. With uniform loading τ^{0}_{ij} applied to the infinite medium.

Table 2.1: Orientation of first slip system of the eight crystals with respect to specimen axes.

Crystal Number	Slip Axes	x_1	x_2	x_3
1	α	-0.1900000	-0.6811025	0.7071068
	β	0.1900000	0.6811025	0.7071068
	γ	0.9632244	-0.2687005	0.0
2	α	-0.1900000	0.7071068	-0.6811025
	β	0.1900000	0.7071068	0.6811025
	γ	0.9632244	0.0	-0.2687005
3	α	0.7071068	-0.6811025	-0.1900000
	β	0.7071068	0.6811025	0.1900000
	γ	0.0	-0.2687005	0.9632244
4	α	0.5773502	0.5773502	0.5773502
	β	0.0	-0.7071068	0.7071068
	γ	0.8164965	-0.4082482	-0.4082482
5	α	0.5773502	0.5773502	0.5773502
	β	0.0	0.7071068	0.7071068
	γ	0.8164965	-0.4082482	-0.4082482
6	α	0.7071068	-0.1900000	-0.6811025
	β	0.7071068	0.1900000	0.6811025
	γ	0.0	0.9632244	0.2687005
7	α	-0.6811025	0.7072068	-0.1900000
	β	0.6811025	0.7071068	0.1900000
	γ	-0.2687005	0.0	0.9632244
8	α	-0.6811025	-0.1900000	0.7071068
	β	0.6811025	0.1900000	0.7071068
	γ	-0.2687005	0.9632244	0.0

α = normal to the slip plane.
β = along the slip direction.
γ = $\perp$ to both a and b.

$$\tau_{ij}(p) = \tau_{ij}^0 + a_{ijqn}(p)\varepsilon^P(q, n). \tag{2.6}$$

Here, the average microstress $\overline{\tau_{ij}}$ over the cubic block of 64 crystals is taken to represent the macroscopic stress S_{ij}. This is different from the self-consistent method, in which the applied stress τ_{ij}^0 is taken to represent the macroscopic stress

$$\Delta S_{ij} = \Delta\overline{\tau_{ij}} = \Delta\tau_{ij}^0 + \overline{a}_{ijqn}\Delta\varepsilon^P(q, n) \tag{2.7}$$

where the bar over a letter denotes the average value.

$$\Delta\tau_{ij}(p) = \Delta S_{ij} + (a_{ijqn} - \overline{a}_{ijqn})\Delta\varepsilon^P(q, k)$$

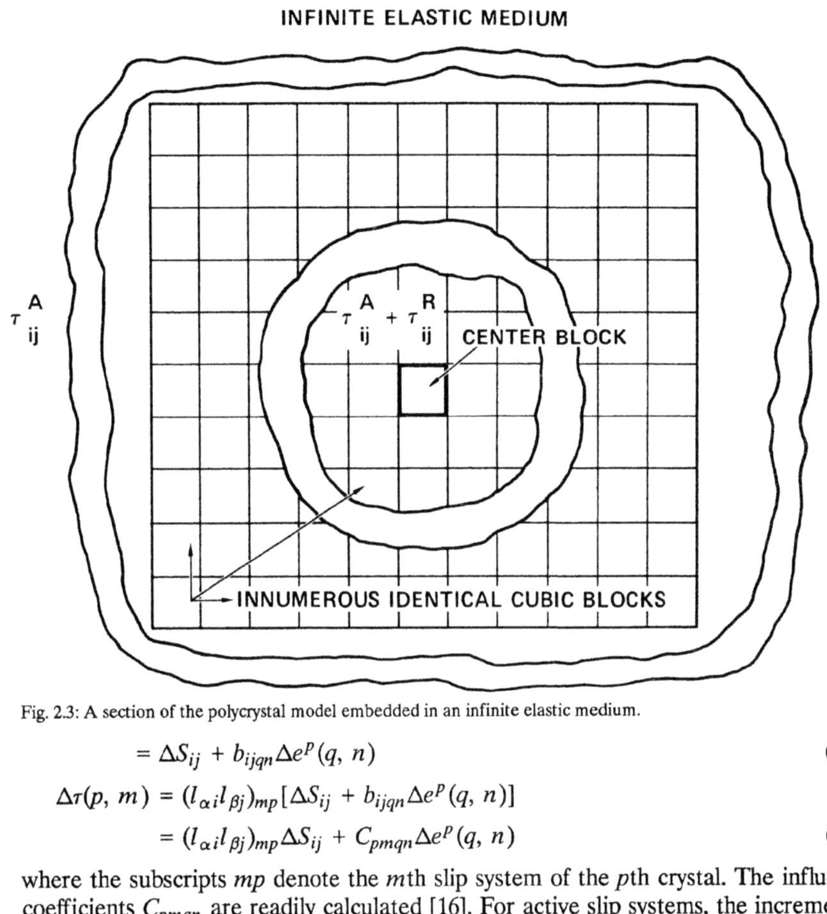

Fig. 2.3: A section of the polycrystal model embedded in an infinite elastic medium.

$$= \Delta S_{ij} + b_{ijqn} \Delta e^{p}(q, n) \tag{2.8}$$

$$\Delta \tau(p, m) = (l_{\alpha i} l_{\beta j})_{mp} [\Delta S_{ij} + b_{ijqn} \Delta e^{p}(q, n)]$$

$$= (l_{\alpha i} l_{\beta j})_{mp} \Delta S_{ij} + C_{pmqn} \Delta e^{p}(q, n) \tag{2.9}$$

where the subscripts mp denote the mth slip system of the pth crystal. The influence coefficients C_{pmqn} are readily calculated [16]. For active slip systems, the incremental resolved shear stress $\Delta\tau(p, m)$ is equated to its incremental critical shear stress $\Delta\tau_c(p, m)$ and all the incremental plastic strains are obtained.

The stress-strain curve of crystals varies with grain size. To take care of the grain size effect, the component crystal stress-strain relation is derived from the experimental tensile stress-strain curve of the polycrystal. This approach is similar to the derivation of the characteristic shear function from the polycrystal tensile stress-strain curve in the development of the first slip theory of plasticity by Batdorf and Budiansky [17]. The polycrystal stress-strain curve of an aluminum alloy 14ST4 is shown in Figure 2.4. The calculated crystal resolved shear stress vs. sum of slip is shown in Figure 2.5. It may be of interest to note that the calculated critical shear stress of component crystal remains constant at 14071 p.s.i in the wide range of the sum of slip from 384×10^{-6} to 7863×10^{-6}. If we use this value as the initial yield stress with no strain-hardening, we found that the polycrystal theoretical self-consistent macroscopic stress-strain curve, calculated by Kroner, Budiansky and Wu's model of spherical crystals, agrees well with the present theory, which considers crystal of cubic shape [14].

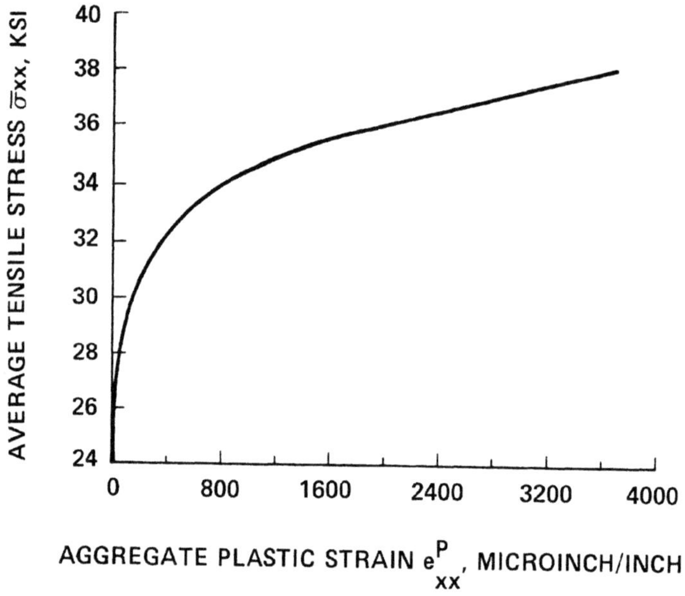

Fig. 2.4: Polycrystal aluminum stress-plastic strain curve in uniaxial tension.

Budiansky *et al.* [18] have experimentally obtained the incremental compression and tension loadings after the material was stressed beyond yielding in compression. The incremental stress-strain relations under these loadings calculated previously by a 27-crystal model and then by the present model and are shown in Figures 2.6 and 2.7. It seems that the present model gives better agreement with the experimental results. Furthermore, as stated previously, the 27-crystal model does not satisfy the conditions given by Equations (2.1) and (2.2).

2.5 Conclusions

This physical theory seems to be the only theory satisfying both the equilibrium and compatibility conditions. Before the development of this 64-crystal model, the condition of initial isotropy was not satisfied. This is a big drawback for the physical theory. With this 64-crystal model, this difficulty is removed. This makes the physical theory fit for practical applications. Numerical results based on this model are shown with those results calculated from a previous model.

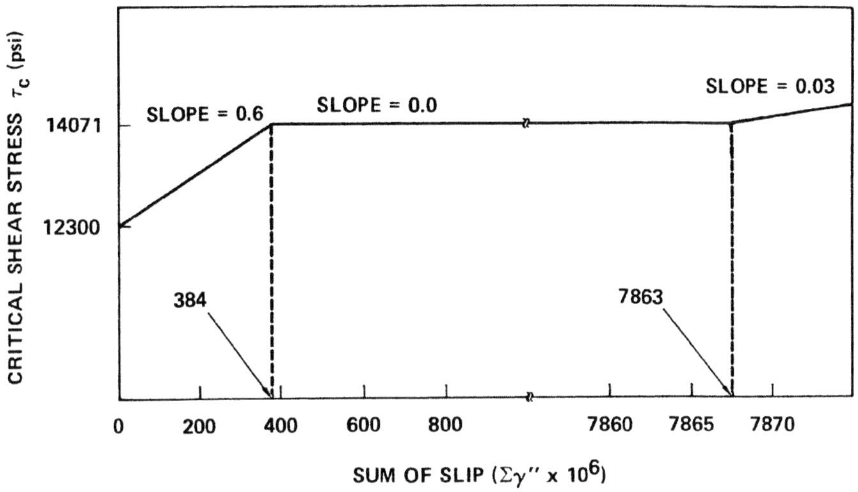

Fig. 2.5: Critical shear stress vs. sum of slip of the component crystal.

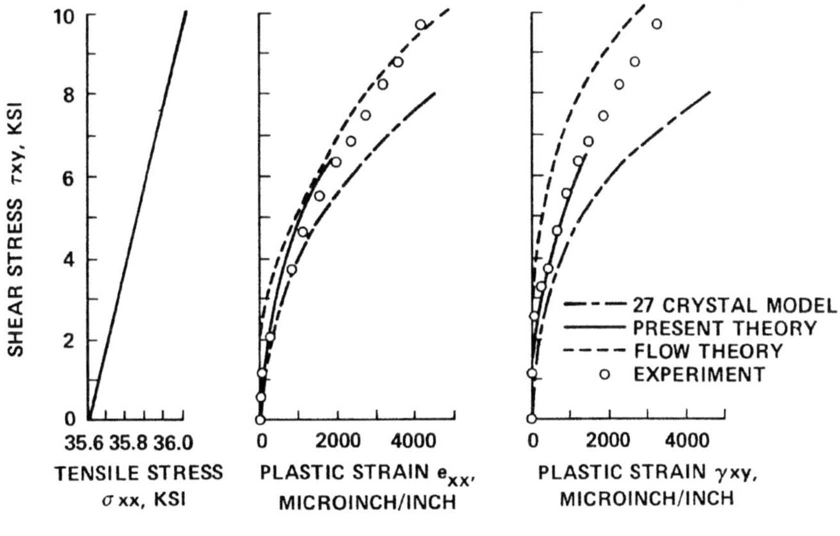

Fig. 2.6: Loading path and plastic strains for $\Delta\sigma_{xx}/\Delta\tau xy = 0.052$.

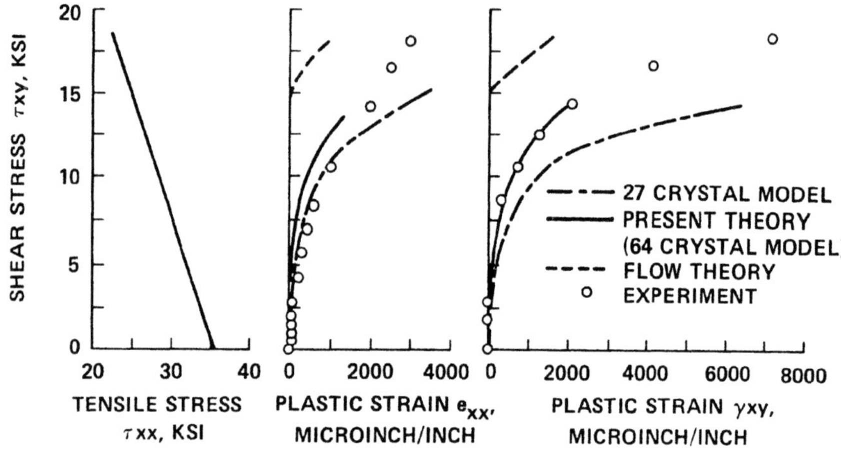

Fig. 2.7: Loading path and plastic strains for $\Delta\sigma_{xx}/\Delta\tau xy = -0.656$.

References

[1] SCHMID, E. and BOAS, W., *Plasticity of Crystals,* Hughes London, 1960, pp. 125 – 152.

[2] TAYLOR, G.I., Plastic Strain in Metals, *J. Inst. Metals,* Vol. 62(1), 1938, pp. 307 – 324.

[3] LIN, T.H., Analysis of Elastic and Plastic Strains of a Face-Centered Cubic Crystal, *J. Mech. Phys. Solids,* Vol. 5, pp. 143 – 159.

[4] TOKUDA, M., KRATOCHVIL, J., and OHASHI, Y., Mechanism of Induced Plastic Anisotropy of Polycrystalline Metals, *Phys. Stat. Sol.,* Vol. 68, 1981, p. 129.

[5] ESHELBY, J.D., The Determination of the Elastic Field of an Ellipsoidal Inclusion and Related Problems, *Proc. Roy. Soc. Ser. A,* 1957.

[6] KRONER, E., Zür plastischen Verformung des Vielkristalls, *Acta. Met.,* Vol. 9, 1961, p. 155.

[7] BUDIANSKY, B. and WU, T.Y., Theoretical Prediction of Plastic Strains of Polycrystals, *Proc. 1st U.S. Nat. Congr. Appl. Mech., 4th,* 1962, p. 1175.

[8] HERSHEY, A.V., The Plasticity of an Isotropic Aggregate of Anisotropic Face-Centered Cubic Crystals, *Journal of Applied Mechanics,* Vol. 21, 1954, pp. 241 – 249.

[9] HILL, R., Continuum Micro-Mechanics of Elastoplastic Polycrystals, *J. Mech. Phys. Solids,* Vol. 14, 1965, pp. 99 – 102.

[10] HUTCHINSON, J.W., Elastic-Plastic Behavior of Polycrystalline Metals and Composites, *Proc. Roy. Soc.* (Ser.A), Vol. 319, 1970, pp. 779 – 784.

[11] WENG, G.J., Self-Consistent Determination of Time-Dependent Behavior of Metals, ASME, *Journal of Applied Mechanics,* Vol. 48, 1981, pp. 41 – 46.

[12] LIN, T.H., UCHIYAMA, S., and MARTIN, D., Stress Field in Metals at Initial Stage of Plastic Deformation, *J. Phys. Mech. Solids,* Vol. 9, 1961, pp. 200 - 209.

[13] LIN, T.H., Physical Theory of Plasticity, *Adv. in Applied Mechanics,* Vol. 11, 1971, pp. 255 - 311.

[14] LIN, T.H., A Physical Theory of Plasticity and Creep, *Journal of Engineering Materials and Technology,* Nov. 1984.

[15] LIN, T.H., *Theory of Inelastic Structures,* pp. 32 and 51 - 54. Wiley, New York, 1968.

[16] LIN, T.H., and RIBEIRO, G.E., Development of a Physical Theory of Plasticity, *Int. J. Solids & Structures,* Vol. 17, 1981, pp. 545 - 551.

[17] BATDORF, S.B. and BUDIANSKY, B., A Mathematical Theory of Plasticity Based on Slip, *Nat. Adv. Comm. Aeronaut, Tech. Notes* 1871.

[18] BUDIANSKY, B., DOW, N.F., PETERS, R.W., and SHEPHERD, R.P., Experimental Studies of Polyaxial Stress-Strain Laws of Plasticity, *Proc. 1st U.S. Nat. Congr. Applt. Mech.,* 1951, pp. 503 - 512.

N.N. Malinin

3

Creep theories in metal forming

3.1 Introduction

Solutions to problems in metal forming are usually based on the theory of plasticity assuming that the solid is rigid-ideal-plastic, elastic-ideal-plastic or rigid-plastic-hardening where creep is not taken into account. The influence of creep becomes essential when metal is deformed at high temperatures and stresses even for a relatively short-time duration of deformation.

The effect of viscosity on metal forming has been discussed in the works of Hencky [1] and Ishlinsky [2]. Tarnovsky *et al.* [3] and Pozdeev *et al.* [4] used the hereditary theory to explain the time history of deformation in metal forming. The same was done in the work of Ilukovich *et al.* [5]. Mathematical difficulties, however, are encountered where the hereditary theory is applied to solve problems involving complex geometries and physical nonlinearity. Experiments on creep by Rabotnov [6] showed that the hereditary theories approach is more suited for polymers and concrete than for metals. Creep recovery is found to be lower than the hereditary theories for predicting the loading and unloading of metals and alloys. It appears more expedient to use creep theories supplemented by experimental data for metal forming where the strains are large but the time for creep is short. Malinin [7, 8] has established that creep theory may be used, giving the functional relationship of stresses, strains and their time rates of change. We first consider the simple case of uniaxial tension before generalizing the treatment of multiaxial stress state.

The general theory of Rabotnov [6] considers creep rate ξ^c at a definite temperature such that it depends on the stress σ and the structural parameters q_k:

$$\xi^c = \phi(\sigma, q_1, ..., q_k, ..., q_n). \tag{3.1}$$

The variation of the structural parameters is given by

$$\mathrm{d}q_k = a_k \mathrm{d}\varepsilon^c + b_k \mathrm{d}\sigma + c_k \mathrm{d}t + f_k \mathrm{d}T, \tag{3.2}$$

where a_k, b_k, c_k and f_k are functions of the creep strain ε^c, stress σ, time t, temperature T and $q_1, ..., q_k, ..., q_n$.

Simple model

For a single parameter t, the flow theory may be stated as [9]

$$\xi^c = \phi_1(\sigma, t). \tag{3.3}$$

A simple analytical dependence of creep rate on stress and time is [10]

$$\xi^c = \sigma^n B(t), \tag{3.4}$$

where n is a constant for a given material and temperature with $B(t)$ being a function of time and temperature. When $B(t)$ is constant, say k, the nonlinear viscous solid is obtained. The corresponding constitutive equation is

$$\xi^c = k\sigma^n. \tag{3.5}$$

If the structural parameter corresponds to the creep strain, then a strain hardening theory [9, 11, 12] prevails with

$$\xi^c = \phi_2(\sigma, \varepsilon^c). \tag{3.6}$$

The analytical expression for the creep rate, creep strain and stress is usually given in the form

$$\xi^c (\varepsilon^c)^b = v(\sigma), \tag{3.7}$$

where b is a constant depending on the temperature. Shesteriokov [13] found the necessary condition that $v(\sigma)$ must satisfy:

$$v''(\sigma) - b\frac{[v'(\sigma)]^2}{[(b+1)v(\sigma)]} > 0 \tag{3.8}$$

which follows from constant stress creep experiments where the accumulation of creep strains increases more rapidly than that predicted by a linear relation. Consider

$$v(\sigma) = a\sigma^n, \tag{3.9}$$

where a and n are constants depending on the temperature and $n > b + 1$. Equations (3.7) and (3.9) give

$$\xi^c (\varepsilon^c)^b = a\sigma^n \tag{3.10}$$

which was in the creep calculations [6, 8] for structural elements. Equation (3.10) reduces to Equation (3.5) for $b = 0$ and coincides with the constitutive equation in the theory of superplasticity [14, 15]:

$$\sigma = A(\xi^c)^{m_1}(\varepsilon^c)^{m_2}, \tag{3.11}$$

where $A = 1/a^{1/n}$, $m_1 = 1/n$ and $m_2 = b/n$. Since $n > b + 1$, it follows that $m_1 + m_2 < 1$. This inequality is not mentioned in the theory of superplasticity. Values of A, m_1 and m_2 in Equation (3.11) can be found in [16, 17].

Multiaxial stress state

A potential θ may be postulated [6] to define the creep rates

$$\xi_{ij}^c = \lambda\frac{\partial \theta}{\partial \sigma_{ij}}. \tag{3.12}$$

If $\theta = \sigma_e^2$ with

$$\sigma_e = \sqrt{\frac{3}{2} S_{ij} S_{ij}} \tag{3.13}$$

being the effective stress and S_{ij} the deviatoric stress, then the following dependence of creep rates on the stresses can be written

$$\xi_{ij}^c = \frac{3\xi_e^c(\sigma_{ij} - \delta_{ij}\sigma_0)}{(2\sigma_e)}, \tag{3.14}$$

where volume change due to creep is neglected and

$$\xi_e^c = \sqrt{\frac{2\xi_{ij}^c \xi_{ij}^c}{3}} \tag{3.15}$$

is the effective creep rate, whose relation to the effective stress in Equation (3.13) depends on the particular creep theory. For uniaxial extension, σ_e is equal to the largest principal stress and ξ_e^c is equal to the largest principal creep rate for an incompressible solid. Hence, the Odqvist's parameter

$$P = \int \mathrm{d}\varepsilon^c = \int \xi^c \,\mathrm{d}t \tag{3.16}$$

is equal to the largest principal strain.

It follows that

$$\xi_e^c = \sigma^n B(t), \tag{3.17}$$

and, for a nonlinear solid:

$$\xi_e^c = k\sigma^n. \tag{3.18}$$

The strain-hardening theory in Equation (3.11) becomes

$$\sigma_e = A(\xi_e^c)^{m_1} P^{m_2}. \tag{3.19}$$

Equations (3.13) – (3.19), Hooke's law, the basic flow theory equation [7], equilibrium equations without acceleration and the strain compatibility conditions can be applied to solve the creep problem, the solutuions of which can be found in [6, 8, 10]. Peculiarities arising from small strain and short-time creep have been discussed in [18]. Hot metal forming creep is also short-time but the strains are large. Investigations on large strains were carried out in the N.E. Bauman Department of 'Strength of Materials and Dynamics and Strength of Machines' at the Moscow Higher Technical School. The results can be found in [19 – 23].

The above investigations showed that, in some cases, the first part of the creep curve is often absent; the equations for a nonlinear viscous solid given by Equation (3.18) can be used. Models of linear and nonlinear viscous solids in metal forming can be found in [15, 24 – 27].

When the initial portion of the creep curves are present, Equations (3.17) and (3.19) can be used. Elastic and plastic strain can be neglected in comparison with creep strains. Henceforth, Equations (3.14) to (3.19) can be referred to as total strain rates and total strains such that the subscript c can be omitted.

3.2 One-dimensional problems

Solutions to three one-dimensional problems are given. The first corresponds to the bending of a sheet under planar strain using a nonlinear viscous model in Equation (3.18) and a strain-hardening theory in Equation (3.19) [28]. Next, a narrow rectangular membrane is solved by application of the strain-hardening theory for the cases of constrained and unconstrained deformation [29, 30]. The third problem deals with a thin walled circular cylindrical tube deformed in a rigid conic die. The nonlinear viscous model is used [31].

Sheet bending under plane strain

The inner and outer surfaces of a sheet are circular cylindrical surfaces with inner radius r_1 and outer radius r_2. The equivalent stress on the circumference at r is

$$\sigma_e = \pm\frac{\sqrt{3}(\sigma_t - \sigma_r)}{2} = A\left[\frac{\dot{\alpha}}{(\sqrt{3}\alpha)}\right]^{m_1}\left[\pm\left(\frac{1 - r_b^2}{r^2}\right)\right]^{m_1} P^{m_2}, \tag{3.20}$$

where σ_t and σ_r are, respectively, the circumferential and radial stresses. The upper sign applies to tension $r_b \leq r \leq r_2$ in the circumferential direction and the lower sign to compression for $r_1 \leq r \leq r_b$ where α is the central angle of a bent sheet, $\dot{\alpha}$ the velocity, and r_b the radius that separates the regions of tension and compression. Radial stresses in the compression and tensile regions are

$$\sigma_r = \frac{2}{\sqrt{3}}\int_{r_1}^{r}\frac{\sigma\,\mathrm{d}r}{r}, \quad \sigma_r = -\frac{2}{\sqrt{3}}\int_{r}^{r_2}\frac{\sigma_e\,\mathrm{d}r}{r}. \tag{3.21}$$

The radius r_b is found by equating the above expressions. The bending moment is equal to:

$$M = \frac{1}{\sqrt{3}}\left(\int_{r_b}^{r_2}\sigma_e r\,\mathrm{d}r - \int_{r_1}^{r_b}\sigma_e r\,\mathrm{d}r\right). \tag{3.22}$$

The velocity of the sheet thickness change is

$$\dot{h} = -\dot{\alpha}(r_2 - r_1)\frac{[1 - r_b^2/(r_1 r_2)]}{(2\alpha)}. \tag{3.23}$$

The radii of curvature r^1 and r in two deformed states are connected by $r^1 = (1 + \xi_t\,\mathrm{d}t)r/(1 + \dot{\alpha}\,\mathrm{d}t/\alpha)$, where $\xi_t = \dot{\alpha}(1 - r_b^2/r^2)/(2\alpha)$ is the circumferential strain rates. Figure 3.1 gives the variations of the radial (curve 1), circumferential (curve 2) stresses and the Odqvist's parameter (curve 3) for a sheet with the following properties: $m_1 = 0.135$ and $m_2 = 0.164$ for the logarithmic circumferential strain on the inner surface with $|\overline{\varepsilon_t}| = 36\%$ at the velocity of $\dot{\alpha} = 1/64$ rad/sec. Figure 3.2

shows the dependence of the bending moment (curve 1), the border surface displacement (curve 2) and the change of a sheet thickness (curve 3) on the Odqvist's parameter at the inner surface.

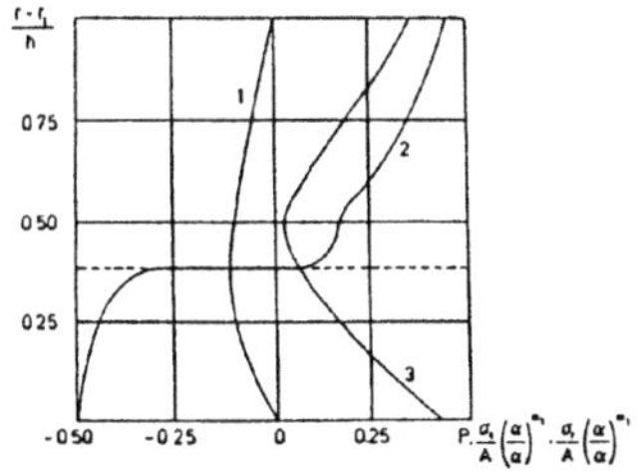

Fig. 3.1: Radial stress (curve 1), circumferential stress (curve 2) and Odqvist's parameter (curve 3) for the bending of a sheet.

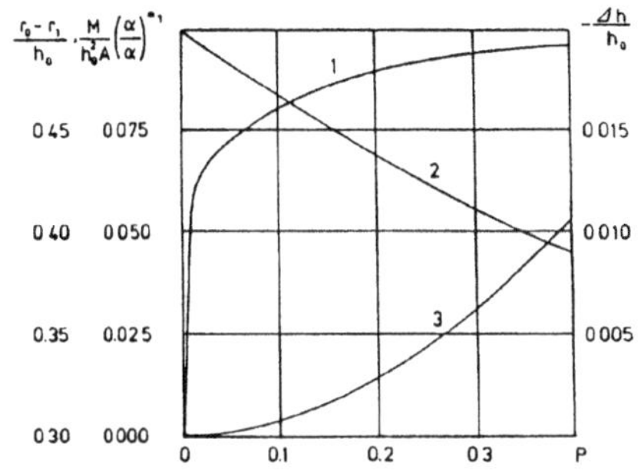

Fig. 3.2: Change of bending moment (curve 1), surface displacement (curve 2) and sheet thickness (curve 3) with Odqvist's parameter at the inner surface.

Narrow rectangular membrane

A narrow rectangular membrane is deformed by a uniform pressure p that varies with time. The membrane has a width $2l$ and thickness h_0, the edges of which are fixed, Figure 3.3.

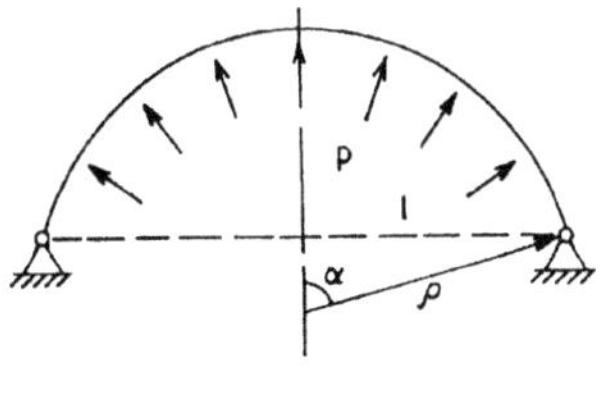

Fig. 3.3: Narrow rectangular membrane in deformed state.

Plane strain is assumed with the axial strain being zero. The middle membrane surface in the deformed state conforms to the circular cylindrical surface with radius ρ and the angle α as indicated in Figure 3.3. The quantities σ_e, ξ_e and P are

$$\sigma_e = \frac{\sqrt{3}\sigma_t}{2} = \frac{\sqrt{3}p\rho}{2h},$$
$$\xi_e = \frac{2(1/\alpha - \operatorname{cotan}\alpha)\dot{\alpha}}{\sqrt{3}}, \tag{3.24}$$
$$P = \frac{2\ln(\alpha/\sin\alpha)}{\sqrt{3}}.$$

The membrane thickness in a deformed state is

$$h = \frac{h_0 \sin\alpha}{\alpha}. \tag{3.25}$$

It follows from Equation (3.25) that the viscous rupture of a membrane ($h = 0$) occurs at $\alpha = \pi$. The equation for determining α at t is

$$\int_0^t p^{1/m_1}\,\mathrm{d}t = c\int_0^\alpha \Psi\,\mathrm{d}\alpha, \tag{3.26}$$

where

$$\Psi = \left(\frac{\sin^2\alpha}{\alpha}\right)^{1/m_1}\left[\ln\left(\frac{\alpha}{\sin\alpha}\right)\right]^{m_2/m_1}\left(\frac{1}{\alpha} - \operatorname{cotan}\alpha\right),$$
$$c = \left(\frac{2}{\sqrt{3}}\right)^{(m_1+m_2+1)/m_1}\left(\frac{Ah_0}{l}\right)^{1/m_1}. \tag{3.27}$$

Once α is known, Equation (3.25) can be used to find the membrane thickness in the deformed state and, hence, the equivalent stress follows from Equation (3.24). Figure 3.4 displays the variations of $\tilde{t} = p^{1/m_1}c^{-1}t$ obtained from Equation (3.26) for p = const with α for the case of a steel membrane at 1150°C with $\underline{A}$ = 145 MPa $\sec^{m_1}$, m_1 = 0.215 and m_2 = 0.165. The viscous rupture at time $\alpha = \pi$ is $\tilde{t}$ = 0.0384.

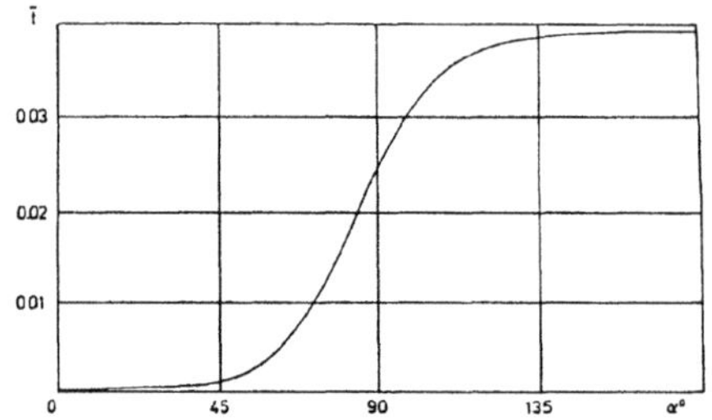

Fig. 3.4: Dependence of angle α on normalized time.

Consider the membrane deformation constrained in a wedge die in Figure 3.5. The membrane will come into contact with the die in such a way that the formation is no longer free. Two limiting boundary conditions at the contact surfaces are considered. They are slipping without friction, and the other is where the membrane sticks to the

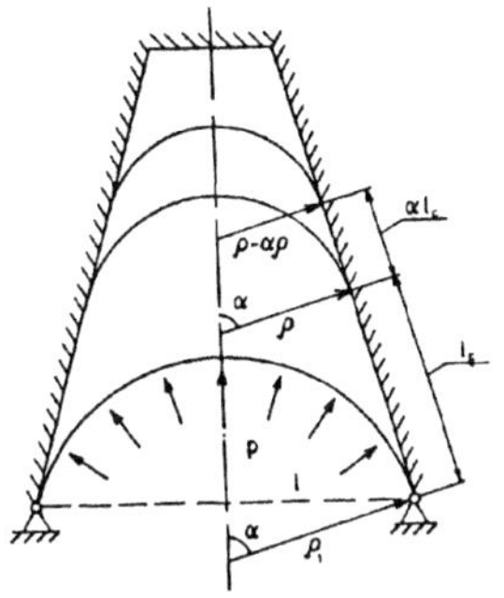

Fig. 3.5: Narrow rectangular membrane in wedge die.

die. The former is considered first. Free deformation terminates at $t = t_1$ and $\rho_1 = l/\sin\alpha$. The length of the contact is denoted by l_c so that $l_c = 0$ at $t = t_1$. The equivalent stress, strain rate and Odqvist's parameter are

$$\sigma_e = \frac{\sqrt{3}\sigma_t}{2} = \left(\frac{\sqrt{3}}{2}\right)\left(\frac{pl}{h_0}\right)(1 - y\cos\alpha) \times \frac{[\alpha + y(1 - \alpha\,\mathrm{cotan}\,\alpha)\sin\alpha]}{\sin^2\alpha},$$

$$\xi_e = \frac{2}{\sqrt{3}}\frac{(1 - \alpha\,\mathrm{cotan}\,\alpha)\dot{y}}{\sqrt{3}\alpha/\sin\alpha + y(1 - \alpha\,\mathrm{cotan}\,\alpha)}, \tag{3.28}$$

$$P = \frac{2\ln[\alpha/\sin\alpha + y(1 - \alpha\,\mathrm{cotan}\,\alpha)]}{\sqrt{3}},$$

where $y = l_c/l, \dot{y} = \mathrm{d}y/\mathrm{d}t$. The membrane thickness in the deformed state is

$$h = \frac{h_0 l}{(\rho\alpha + l_c)} = h_0\left[\frac{\alpha}{\sin\alpha + y(1 - \alpha\,\mathrm{cotan}\,\alpha)}\right]^{-1}. \tag{3.29}$$

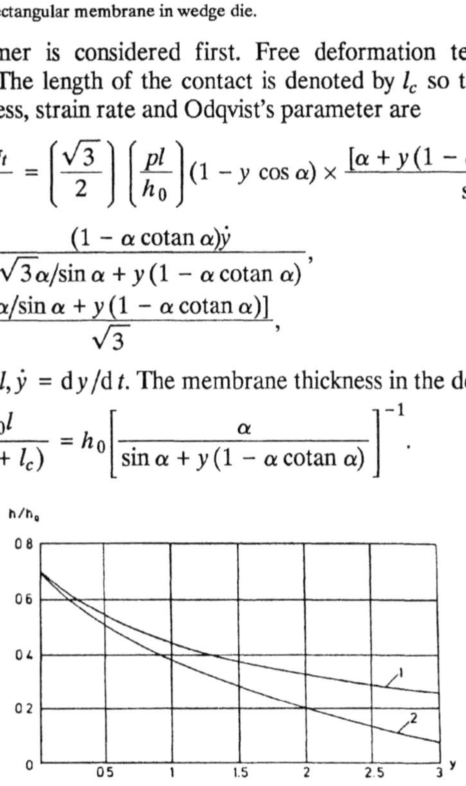

Fig. 3.6: Dependence of membrane thickness on normalized contact length for slipping (curve 1) and sticking (curve 2).

Figure 3.6 shows the variations of h/h_0 with y for $\alpha = 80^\circ$ (curve 1) where $h = h_1 = h_0 \sin\alpha/\alpha$. The contact length is

$$\int_{t_1}^{t} P^{1/m_1}\,\mathrm{d}t = c\int_0^y \Psi_1\,\mathrm{d}y, \tag{3.30}$$

in which

$$\Psi_1 = \frac{\kappa_1}{\kappa_2 + \kappa_1 y}[\ln(\kappa_2 + \kappa_1 y)]^{m_2/m_1}\left\{\frac{\alpha}{[\kappa_2 - (1 - \kappa_1)(\kappa_2 + \kappa_1 y)]\,y}\right\}^{1/m_1},$$

$$\kappa_1 = 1 - \alpha\,\mathrm{cotan}\,\alpha,\ \kappa_2 = \frac{\alpha}{\sin\alpha}. \tag{3.31}$$

Equation (3.26) gives the α and t_1 corresponding to free deformation. The results of Equation (3.30) for $\alpha = 80^\circ$ are given in Figure 3.7 where the pressure was held constant during the deformation.

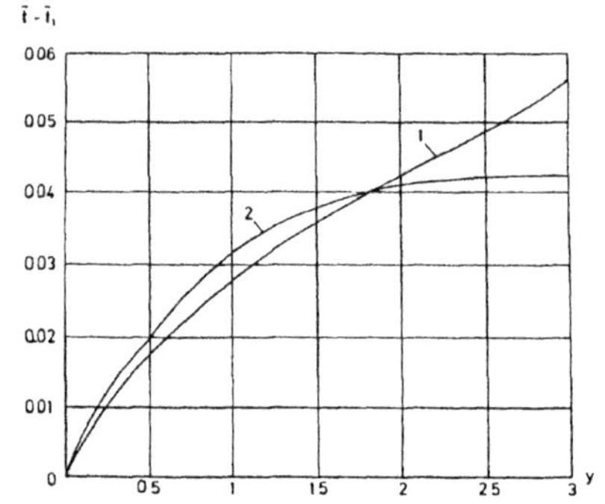

Fig. 3.7: Dependence of normalized time on normalized contact length for slipping (curve 1) and sticking (curve 2).

If the membrane sticks to the die, its thickness in the part l_c will vary. This problem was solved in [32] for a nonlinear viscous material. The equivalent stress, strain rate and the Odqvist's parameter are determined by

$$\sigma_e = \frac{\sqrt{3}\sigma_t}{2} = \frac{(\sqrt{3}/2)(pl/h_0)(1 - y\cos\alpha)^{1-\kappa}\alpha}{\sin^2\alpha},$$

$$\xi_e = \frac{2\kappa\dot{y}\cos\alpha}{[\sqrt{3}(1 - y\cos\alpha)]}, \tag{3.32}$$

$$P = -\left(\frac{2}{\sqrt{3}}\right)\ln\left[\frac{\sin\alpha(1 - y\cos\alpha)^{\kappa}}{\alpha}\right],$$

where $\kappa = \alpha^{-1}\tan\alpha - 1$. The free part of the membrane has thickness $h = h_1(\rho/\rho_1)^{\kappa} = h_0\sin\alpha(1 - y\cos\alpha)^{\kappa}/\alpha$. Curve 2 in Figure 3.6 gives the dependence of h/h_0 on y for $\alpha = 80^\circ$. Curves 1 and 2 reveal that the thickness of the free part is smaller under the sticking condition as compared with the sliding one. The equation for contact length is

$$\int_{t_1}^{t} P^{1/m_1}\, \mathrm{d}t = c\int_{0}^{y} \Psi_2\, \mathrm{d}y, \tag{3.33}$$

where

$$\Psi_2 = \frac{\kappa \cos\alpha}{1 - y\cos\alpha}\left[\ln \frac{\alpha}{\sin\alpha(1 - y\cos\alpha)^{\kappa}}\right]^{m_2/m_1}\left[\frac{\sin^2\alpha}{\alpha(1 - y\cos\alpha)^{1-\kappa}}\right]^{1/m_1}. \tag{3.34}$$

The time dependent nature of Equation (3.33) can be seen in Figure 3.7 for $\alpha = 80^{\circ}$.

Tube drawing

Consider the deformation of a thin-walled circular cylindrical tube with a constant cross section in the rigid conical die. The different cases are shown in Figures 3.8(a) to 3.8(d). The bending moment will be neglected and the Coulomb's law of friction will be used. The tube deformation is axisymmetric. Therefore, stresses, strains and wall thickness depend only upon the radius. The meridional σ_m and circumferential σ_t stresses are

$$\sigma_m = \frac{(2\sigma_e \cos\phi)}{\sqrt{3}}, \quad \sigma_t = \frac{[2\sigma_e \cos(\phi - \pi/3)]}{\sqrt{3}}, \tag{3.35}$$

with

$$\sigma_e = \sqrt{\sigma_m^2 - \sigma_m\sigma_t + \sigma_t^2}. \tag{3.36}$$

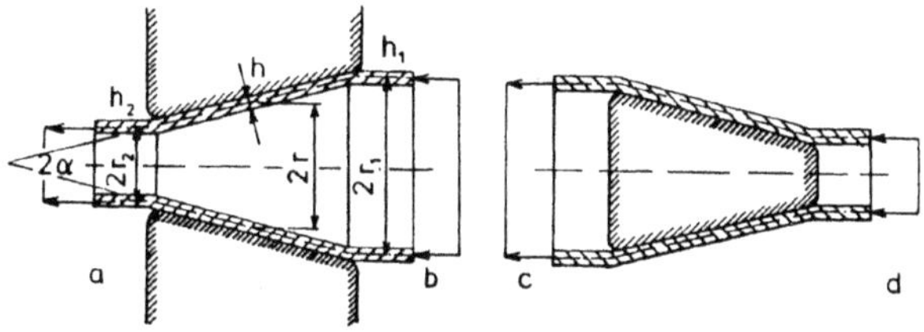

Fig. 3.8: Different cases of tube deformation in a rigid conical die.

being the effective stress. Limits to the change of ϕ can be established for the tube drawing in Figure 3.8(a) $\sigma_m > 0$, $\sigma_t < 0$ for $3\pi/2 \leq \phi \leq 11\pi/6$; in Figure 3.8(b) $\sigma_m < 0$, $\sigma_t < 0$ for $5\pi/6 \leq \phi \leq 3\pi/2$; in Figure 3.8(c) $\sigma_m > 0$, $\sigma_t > 0$ for $-\pi/6 \leq \phi \leq \pi/2$ and in Figure 3.8(d) $\sigma_m < 0$, $\sigma_t > 0$ for $\pi/2 \leq \phi \leq 5\pi/6$. The solution for the equivalent stress is

$$\sigma_e = a_1\left(\frac{h}{h_1}\right)^{-1/n}\left(\frac{r}{r_1}\right)^{-2/n} |\sin\phi|^{-1/n}$$

$$= a_2\left(\frac{h}{h_2}\right)^{-1/n}\left(\frac{r}{r_2}\right)^{-2/n} |\sin\phi|^{-1/n}, \tag{3.37}$$

where $a_1 = [v_1 \sin\alpha/(\mu r_1)]^{1/n}$, $a_2 = [v_2 \sin\alpha/(\mu r_2)]^{1/n}$, $n = 1/m_1$ and $\mu = 1 + f\tan\alpha$ with f being the coefficient of friction. Figure 3.8 gives the die taper angle 2α, the die radii r_1 and r_2, the tube velocities v_1 and v_2 and the thickness h_1 at the entrance and h_2 at the exit.

The function ϕ is related to the radius r by the equations

$$\frac{r}{r_1} = \exp\int_{\phi_1}^{\phi} Y\phi, \quad \frac{r}{r_2} = \exp\int_{\phi_2}^{\phi} Y\,d\phi \tag{3.38}$$

depending on whether the meridional stress is that at the entrance or exit of the die. The symbol Y represents

$$Y = \frac{2(1 + n\tan^2\phi)}{-\sqrt{3}(n-1) + [n(1-\mu) - 3]\tan\phi - \sqrt{3}\mu n\tan^2\phi}, \tag{3.39}$$

where ϕ_1 and ϕ_2 are the values of the function ϕ at the die entrance and die exit, respectively. They are determined by

$$|\sin\phi_1|^{-1/n}\cos\phi_1 = \frac{\sqrt{3}\sigma_{m_1}}{(2a_1)},$$

$$|\sin\phi_2|^{-1/n}\cos\phi_2 = \frac{\sqrt{3}\sigma_{m_2}}{(2a_2)}. \tag{3.40}$$

The meridional stresses at the opposite end of the die are obtained from Equations (3.35) and (3.37) since the dependence of ϕ and r can be bound from Equations (3.38). The wall thickness of a tube is determined from

$$\frac{h}{h_1} = \exp\left[-\int_{\phi/1}^{\phi}\frac{\sqrt{3}\cos\phi + \sin\phi}{2\sin\phi} Y\,d\phi\right]. \tag{3.41}$$

Plotted in Figure 3.9 are the normalized meridional stress (curve 1), circumferential stress (curve 2) and wall thickness (curve 3) as a function of r/r_1 for the case tube drawing in Figure 3.8(a) with $\alpha = 14°$, $f = 0.1$ and $n = 3{,}6$. The maximum possible reduction in a tube corresponds to $\sigma_t \rightarrow 0$.

3.3. Reduction from two to one dimension

Two assumptions are made in calculation the plastic deformation of metal sheets [33 – 35]. They are that plane sections remain plane, and the uniformity of stresses and strains across the width. The former is invalidated by the presence of shear stress at the contact points whereas incompressibility leads to changes in strain. The latter reduces a two-dimensional problem to that of one-dimension. These simplifications made it possible to obtain closed solutions to a number of problems, such as:

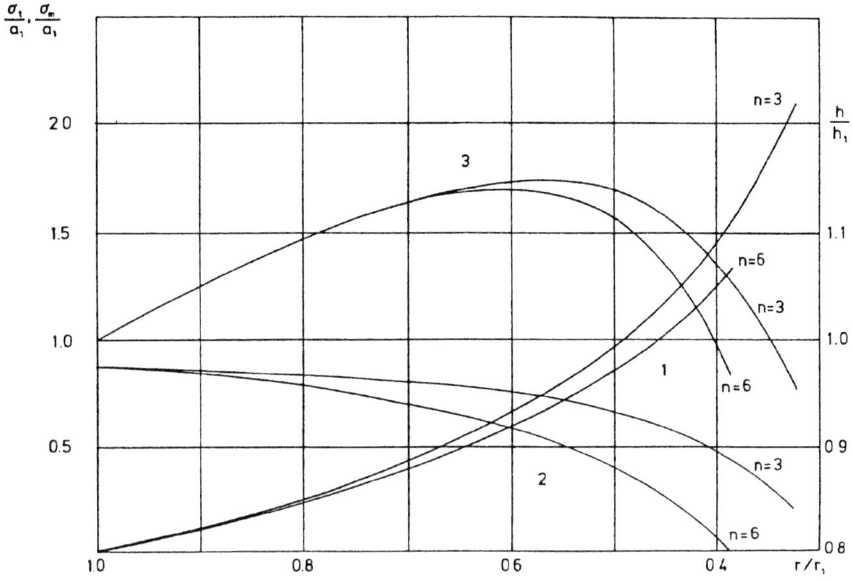

Fig. 3.9: Meridional stress (curve 1) circumferential stress (curve 2) and wall thickness (curve 3).

compression of a bar under plane strain; compression of a solid circular cylinder; extrusion of a rod through a conical die [36]; and free and fixed compression of a hollow circular cylinder [37].

Based on the constitutive relations of the creep theories in the work, forming loads can be obtained, dependent on the velocity in a given process. Plasticity theory is not able to yield this information. Results for the problems mentioned earlier will be given and they refer to three zones of contact: the slip zone within which Coulomb's law of friction is valid; the brake zone where the frictional force intensity is equal to that of the maximum shear stress; and the stick zone where the intensity of the frictional force passes through zero and behaves linearly. These conditions were used to solve a variety of problems that can be found in [33 – 35, 38, 39].

Fixed deformation of solid circular cylinder

Let a cylinder with diameter d_0 and length $2h_0$ be compressed by two end plates that move towards each other with a velocity v. The effective stress constant in the cylinder is

$$\sigma_e = A\left[\frac{v}{(2h)}\right]^{m_1}\left[\ln\left(\frac{h_0}{h}\right)\right]^{m_2}, \tag{3.42}$$

where $2h$ is deformed height of the cylinder. The contact pressure p and the intensity of frictional forces q in the slip zone at $r_{b_1} \leq r \leq d/2$ is

$$p = \sigma_e \exp\left[\frac{f(\mathcal{D}/2 - r)}{h}\right], \quad q = fp, \tag{3.43}$$

in which f is the coefficient of friction and d the deformed diameter. The radius of the slip and brake zone border is determined by:

$$\frac{(\mathcal{D}/2 - r_{b_1})}{h} = -\frac{(\ln 2f)}{f}. \tag{3.44}$$

In the brake zone at $r_{b_2} = 2h \leq r \leq r_{b_1}$, it turns out that

$$p = p_1 + \frac{\sigma_e(r_{b_1} - r)}{(2h)}, \quad q = \frac{\sigma_e}{2}, \tag{3.45}$$

in which P_1 is given by Equation (3.43) with $r = r_{b_1}$. In the stick zone at $0 \leq r \leq r_{b_2} = 2h$, it is found that

$$p = p_2 + \frac{\sigma_e[1 - r^2/(4h^2)]}{2}, \quad q = \frac{\sigma_e r}{(4h)}, \tag{3.46}$$

in which P_2 is given by Equation (3.45) with $r = r_{b_2} = 2h$.

Free deformation of hollow circular cylinder

Consider a hollow cylinder with external diameter d_0 and internal diameter d_1. For the case under consideration, a bilateral metal flow can occur; an inward and an outward flow can exist. A neutral surface with radius r_n prevails on the radial displacements have zero velocity. The circumferential σ_t and axial σ_z stresses are connected with the radial stress σ_r by

$$\sigma_t = \sigma_r - 2x^2\chi, \quad \sigma_z = \sigma_r - (x^2 + 3)\chi, \tag{3.47}$$

where

$$\chi = g(x^4 + 3)^{(m_1-1)/2} P^{m_2}/\sqrt{3},$$

$$g = A\left[\frac{v}{(2\sqrt{3}h_0)}\right]^{m_1}\left(\frac{h_0}{h}\right)^{m_1}, \quad x = \frac{r_n}{r},$$

$$P = \frac{1}{\sqrt{3}} \int_h^{h_0} (x^4 + 3)^{1/2} \frac{\mathrm{d}h}{h}. \tag{3.48}$$

The radial stresses for $r_1 \leq r \leq r_n$ are

$$\sigma_r = -2\exp\left(\frac{\beta_1}{x}\right) \int_x^{x_1} \exp\left(-\frac{\beta_1}{x}\right) \chi x \, \mathrm{d}x -$$

$$- \beta_1 \exp\left(\frac{\beta_1}{x}\right) \int_x^{x_1} \exp\left(-\frac{\beta_1}{x}\right) \chi(x^2 + 3)\, x^{-2}\, \mathrm{d}\, x, \tag{3.49}$$

and for r_n , $= r \leq r_2$ they are

$$\sigma_r = 2 \exp\left(-\frac{\beta_1}{x}\right) \int_{x_2}^{x} \exp\left(\frac{\beta_1}{x}\right) \chi x \, \mathrm{d}\, x \; -$$

$$- \beta_1 \exp\left(-\frac{\beta_1}{x}\right) \int_{x_2}^{x} \exp\left(\frac{\beta_1}{x}\right) \chi(x^2 + 3)\, x^{-2}\, \mathrm{d}\, x, \tag{3.50}$$

where $\beta_1 = fr_n/h, x_1 = r_n/r_1$ and $x_2 = r_n/r_2$. The two aforementioned regions can be obtained from Equations (3.49) and (3.50) by letting $x = 1$. To this end, the deformation is divided into small segments so that r_n can be found for known values of r_1 and r_2 corresponding to h. The condition of constant volume is used to obtain r_1^1 and r_2^1 corresponding to h^1 for the subsequent segment, i.e.

$$h(r_2^2 - r_n^2) = h^1[(r_2^1)^2 - r_n^2], \; h(r_n^2 - r_1^2) = h^1[r_n^2 - (r_1^1)^2]. \tag{3.51}$$

The incremental change of the Odqvist parameter ΔP according to Equation (3.48) is

$$\Delta P = \frac{1}{\sqrt{3}} \int_{h'}^{h} (x^4 + 3)^{1/2} \frac{\mathrm{d}h}{h} = \frac{1}{\sqrt{3}} (x^4 + 3)^{1/2} \ln\frac{h}{h'}, \tag{3.52}$$

where x can be regarded as constant during the deformation. By the same procedure, a relation involving r'', r_n'' and h'' $(h'' > h)$ is obtained:

$$h''[(r_n'')^2 - (r'')^2] = h[(r_n'')^2 - r^2] \tag{3.53}$$

where r'' can be either smaller or larger than r_n''. Denoting $x = r_n''/r$ and $x'' = r_n''/r''$, the result is

$$x = \left\{\frac{1 - h''[1 - (x'')^{-2}]}{h}\right\}^{-1/2}. \tag{3.54}$$

The stresses follow from Equations (3.49), (3.50) and (3.47). The foregoing solution is valid if the intensity of the friction forces is lower than the maximum shear stress τ_{max} given by

$$\tau_{max} = \left[\frac{2}{(2 + \sqrt{3})}\right] \sigma_e = 0.535\sigma_e. \tag{3.55}$$

If the intensity of the frictional forces is equal to the maximum shear stress at all points of contact, then

$$q = \tau_{max} = \eta\, \sigma_e \quad (\eta = 0.535). \tag{3.56}$$

In this case, the radial stress for this $r_1 \leq r \leq r_n$ is

$$\sigma_r = -2\int_x^{x_1} \chi x \,\mathrm{d}x - \sqrt{3}\beta_2 \int_x^{x_1} \chi x^{-2}\,\mathrm{d}x, \tag{3.57}$$

and for $r_n \leq r \leq r_2$ is

$$\sigma_r = 2\int_{x_2}^{x} \chi x \,\mathrm{d}x - \sqrt{3}\beta_2 \int_{x_2}^{x} \chi x^{-2}\,\mathrm{d}x, \tag{3.58}$$

where $\beta_2 = \eta r_n / h$.

Numerical results are obtained for $2h_0 = 15$ mm, $d_1 = 5$ mm, $d_0 = 25$ mm and $v = 0.167$ mm/sec while the aluminum cylinder with $A = 0.881 \times 10^{-2}$ Pa/sec^m, $m_1 = 0.2$ and $m_2 = 0$ is heat to a temperature of 450°C. Assuming that $\eta = 0.5$, the calculated diameters d_1 and d in Table 3.1 are compared with those obtained by experiments $d_1^{\exp}$ and $d^{\exp}$ while d_n corresponds to the neutral surface. Figure 3.10 compares the theoretical results (curve 1) with the experiments (curve 2). The discrepancy of the deformation force does not exceed 6% and the deviation of the calculated and measured radius is 11%.

Table 3.1: Comparison of theory and experiment for the free compression of a hollow circular cylinder.

$[(h_0 - h)/h_0]100$	d_1 mm	$d_1^{\exp}$ mm	d mm	$d^{\exp}$ mm	d_n mm
8	4.88	4.94	26.0	26.1	6.76
16	4.64	4.88	27.0	28.0	7.16
20	4.48	4.72	27.8	28.4	7.52
29	3.84	4.32	29.0	30.0	7.42

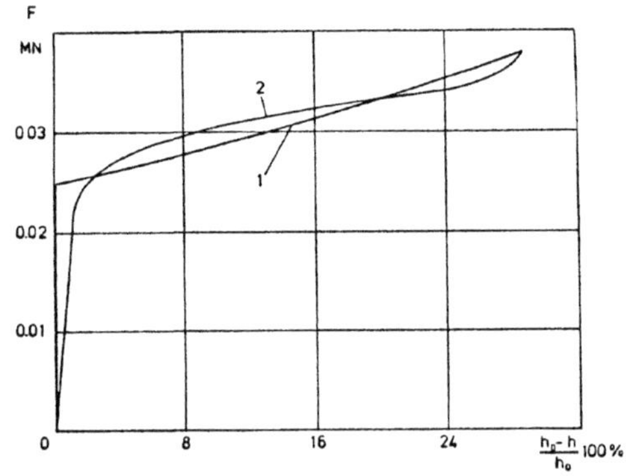

Fig. 3.10: Theoretical (curve 1) and experimental (curve 2) comparison for a compressed hollow cylinder.

Fixed deformation of hollow circular cylinder

A rigid yoke is used to constrain the expansion of the outer diameter of the hollow cylinder. Unlike the previous example, which made use of Equation (3.48), the incompressibility condition is employed to establish the relation

$$h_0(r_2^2 - r_0^2) = h(r_2^2 - r^2). \tag{3.59}$$

It follows that for $x_0 = r_2/r_0$ and $x = r/r_0$, Equation (3.59) becomes

$$x = \left[\frac{1 - h_0(1 - x_0^{-2})}{h}\right]^{-1/2}. \tag{3.60}$$

Making use of Equation (3.60), Equation (3.48) can be expressed as

$$P = \frac{1}{\sqrt{3}} \int_h^{h_0} \left\{3 + \left[1 - \frac{h_0}{h}(1 - x_0^{-2})\right]^{-2}\right\}^{1/2} \frac{dh}{h}. \tag{3.61}$$

If Coulomb's friction law is used, then the equation for the radial stress is

$$\sigma_r = -\exp\left(\frac{\beta_3}{x}\right) \int_x^{x_1} \exp\left(-\frac{\beta_3}{x}\right) x\left[2x + \frac{\beta_3}{x^2}(x^2 + 3)\right] dx, \tag{3.62}$$

where $\beta_3 = fr_2/h$. The upper limit of integration is

$$x_1 = \left[\frac{1 - h_0(1 - x_{10}^{-2})}{h}\right]^{-1/2}, \tag{3.63}$$

where $x_{10} = r_2/r_{10}$.

If the intensity of the frictional force is equal to the maximum shear stress then

$$\sigma_r = -\int_x^{x_1} x\left[2x + \frac{\sqrt{3}\beta_4}{x^2}(x^4 + 3)\right]^{1/2} dx, \tag{3.64}$$

where $\beta_4 = \eta r_2/h$. Again, the circumferential and axial stresses are obtainable from Equation (3.47).

When the contact plane divides $r_1 \leq r \leq r_*$, where Coulomb's friction force applies, and $r_* \leq r \leq r_2$ where the frictional force intensity is equal to the maximum shear stress, then the radial for $r_1 \leq r \leq r_*$ is given by Equation (3.62). The value r_* is determined from the condition $-f\sigma_z = \eta\,\sigma_e$. The radial stress for $r_* \leq r \leq r_2$ is obtained from Equation (3.64):

$$\sigma_r = \sigma_{r_*} - \int_x^{x_*} x\left[2x + \frac{\sqrt{3}\beta_2}{x^2}(x^4 + 3)^{1/2}\right] dx, \tag{3.65}$$

where $\sigma_{r/*}$ corresponds to $r = r_*$ and $x_* = r_2/r_*$.

Numerical results of the radial stress (curve 1), axial stress (curve 2), circumferential stress (curve 3), frictional force (curve 4), and maximum shear (curve 5) are displayed in Figure 3.11 for $(h_0 - h)100/h_0 = 15.2\%$, $r_* = 0.68\,r_2$, $2h_0 = d_0$, $d_0 = 0.4\,d_0$ and $f = 0.1$ while the material and temperature are the same as those in the preceding example.

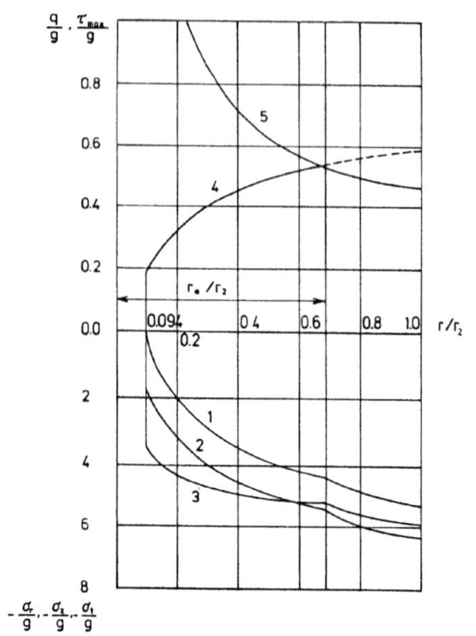

Fig. 3.11: Radial stress (curve 1), axial stress (curve 2), circumferential stress (curve 3), frictional force (curve 4) and maximum shear stress (curve 5) for the compression of a hollow cylinder in a rigid yoke.

Consider the compression of a hollow cylinder which is inserted with a rigid bar. Using the Coulomb's law, the radial stress is given by

$$\sigma_r = \exp\left(-\frac{\beta_5}{x}\right)\int_{x_2}^{x}\exp\left(\frac{\beta_5}{x}\right)\chi\left[2x - \frac{\beta_5}{x^2}(x^2+3)\right]\mathrm{d}x, \tag{3.66}$$

where $\beta_5 = fr_1/h$. The lower limit of integration is

$$x_2 = \left[\frac{1 - h_0(1 - x_{20}^{-2})}{h}\right]^{-1/2}, \tag{3.67}$$

where $x_{20} = r_1/r_{20}$. If the frictional force intensity is equal to the maximum shear stress, then

$$\sigma_r = \int_{x_2}^{x}\chi\left[2x - \frac{\sqrt{3}\beta_6}{x^2}(x^4+3)^{1/2}\right]\mathrm{d}x, \tag{3.68}$$

where $\beta_6 = r_1/h$. Equations (3.66) and (3.68) can be used to yield

$$\sigma_r = \sigma_{r*} + \int_{x_*}^{x}\chi\left[2x - \frac{\sqrt{3}\beta_6}{x^2}(x^4+3)^{1/2}\right]\mathrm{d}x, \tag{3.69}$$

where σ_{r*} corresponds to $r = r_*$ that is determined according to Equation (3.66) with $x = r_1/r_*$.

Shown in Figure 3.12 are the numerical results for the radial stress (curve 1), the axial stress (curve 2), and the circumferential stress (curve 3) for $(h_0 - h)100/h_0 = 15\%$, $2h_0 = 1.5\,d_0$, $d_0 = 5\,d_0$ while the temperature and material are the same as those in the previous example.

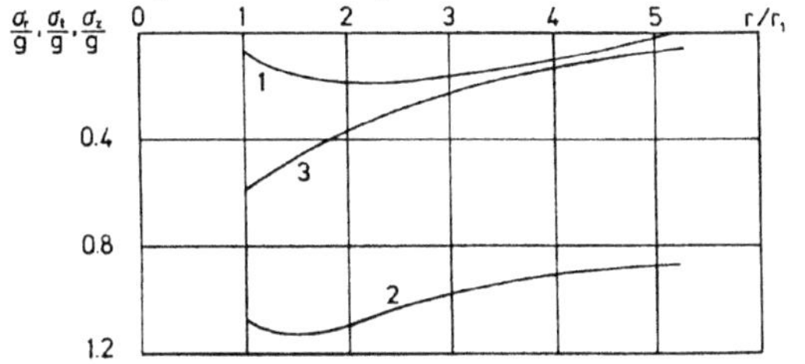

Fig. 3.12: Radial stress (curve 1), axial stress (curve 2), and circumferential stress (curve 3) for the compression of hollow cylinder with an inserted bar.

Rolling of a strip

The rolling of a strip is shown in Figure 3.13. A thickness reduction of h_1 to h_2 is seen. Deformation is assumed to occur between entrance and exit while the remaining strip is considered undeformed.

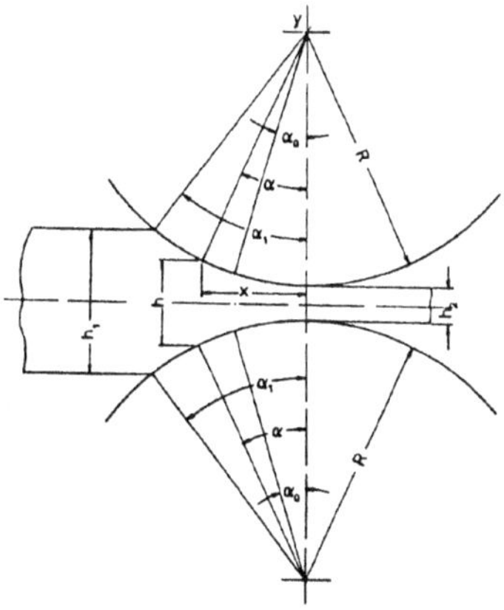

Fig. 3.13: Schematic of strip rolling

If the strip width is approximately five times that of its length, the deformation in the strip can be assumed to be planar. The following condition is then valid:

$$v = \frac{v_1 h_1}{h} = \frac{v_2 h_2}{h}. \tag{3.70}$$

Here, v_1 and v_2 are the mean velocities of the strip at the entrance and exit where the thickness of the strip is h determined by the angle α.

The difference in the mean velocity and circumferential velocity of the rolls can give rise to frictional forces, the angular position at which this force changes sign is α_n referred to as the neutral position which distinguishes the forward from the rearward region. The former refers to the exit and the latter to the entrance. The effective stress formula is

$$\sigma_e = \frac{\sqrt{3}(\sigma_x - \sigma_y)}{2} = \frac{\sqrt{3}(\sigma_x + p)}{2}$$
$$= \gamma\left(\frac{h_1^2 \tan\alpha}{h^2}\right)^{m_1}\left[\ln\left(\frac{h_1}{h}\right)\right]^{m_2}, \tag{3.71}$$

where σ_x is the stress normal to the strip, $\sigma_y \approx -p$ the stress parallel to the strip and

$$\gamma = A\left[\frac{4v_1}{(\sqrt{3}h_1)}\right]^{m_1}\left(\frac{2}{\sqrt{3}}\right)^{m_2}. \tag{3.72}$$

In contrast to the previous example, the effective stress is not constant, it varies with h and α, both of which depend on x. Using Coulomb's law, σ_x in the rearward region is

$$\sigma_x = \exp\left\{-2fR\left[I_1(\alpha) - I_1(\alpha_1)\right]\right\}\left\{\sigma_{x_1} - \left(\frac{4R}{\sqrt{3}}\right)\left[I_2(\alpha) - I_2(\alpha_1)\right]\right\}, \tag{3.73}$$

and in the forward region is

$$\sigma_x = \exp\left[2fRI_1(\alpha)\right]\left[\sigma_{x_2} - \left(\frac{4R}{\sqrt{3}}\right)I_3(\alpha)\right], \tag{3.74}$$

where f is the coefficient of friction while σ_{x_1} and σ_{x_2} are, respectively, the stress at the entrance and exit. Moreover, the following definitions hold:

$$I_1(\alpha) = \int_0^\alpha \frac{\cos\alpha \, d\alpha}{h},$$
$$I_2(\alpha) = \int_0^\alpha \exp\left\{(2fR)\left[I_1(\alpha) - I_1(\alpha_1)\right]\left(\frac{\sigma_e}{h}\right)(\sin\alpha - f\cos\alpha)\, d\alpha,\right. \tag{3.75}$$
$$I_3(\alpha) = \int_0^\alpha \exp\left[-2fRI_1(\alpha)\right]\left(\frac{\sigma_e}{h}\right)(\sin\alpha + f\cos\alpha)\, d\alpha$$

The above solution holds for

$$\tau_{\max} = \frac{(\sigma_x - \sigma_y)}{2} = \frac{(\sigma_x + p)}{2} = \frac{\sigma_e}{\sqrt{3}} \tag{3.76}$$

equal to the frictional force intensity such that

$$fp \le \frac{\sigma_e}{\sqrt{3}} \quad \text{or} \quad \frac{\sigma_x}{\sigma_e} \ge \frac{(2f - 1)}{\sqrt{3}f}. \tag{3.77}$$

The angles $\overline{\alpha}_1$ in the backward region and $\overline{\alpha}_2$ in the forward region can be found from Equations (3.71) – (3.77); they determine the slip zones which are adjacent to the brake zones. In these zones, $q = \tau_{max} = \sigma_e/\sqrt{3}$. The stress σ_x in the rearward and forward regions are given, respectively, by

$$\sigma_x = \overline{\sigma}_{x_1} - \frac{2R}{\sqrt{3}} \int_{\overline{\alpha}_1}^{\alpha} \frac{\sigma_e}{h}(2 \sin \alpha - \cos \alpha) \, \mathrm{d}\, \alpha, \tag{3.78}$$

and

$$\sigma_x = \overline{\sigma}_{x_2} - \frac{2R}{\sqrt{3}} \int_{\overline{\alpha}_2}^{\alpha} \frac{\sigma_e}{h}(2 \sin \alpha + \cos \alpha) \, \mathrm{d}\, \alpha, \tag{3.79}$$

where $\overline{\alpha}_{x_1}$ and $\overline{\alpha}_{x_2}$ are the stresses on the border of the slip and brake zone with angles $\overline{\alpha}_1$ and $\overline{\alpha}_2$. They can be calculated from Equations (3.73) and (3.74). The stress σ_x in the backward region for the condition of sticking is

$$\sigma_x = \tilde{\sigma}_{x_1} - \left(\frac{4R}{\sqrt{3}}\right) [I_4(\alpha) - I_4(\tilde{\alpha}_1)] + \\ + \left(\frac{2\tilde{q}_1 R^2}{l_{s_1}}\right)\left\{[I_5(\alpha) - I_5(\tilde{\alpha}_1)] - \sin \alpha_0 [I_1(\alpha) - I_1(\tilde{\alpha}_1)]\right\}, \tag{3.80}$$

and

$$\sigma_x = \tilde{\sigma}_{x_2} - \left(\frac{4R}{\sqrt{3}}\right) [I_4(\alpha) - I_4(\tilde{\alpha}_2)] - \\ - \left(\frac{2\tilde{q}_2 R^2}{l_{s_2}}\right)\left\{\sin \alpha_0 [I_1(\alpha) - I_1(\tilde{\alpha}_2)] - [I_5(\alpha) - I_5(\tilde{\alpha}_2)]\right\}, \tag{3.81}$$

applies to the forward region where

$$I_4(\alpha) = \int_0^{\alpha} \frac{\sigma_e \sin \alpha \, \mathrm{d}\alpha}{h}, \quad I_5(\alpha) = \int_0^{\alpha} \frac{\sin \alpha \cos \alpha \, \mathrm{d}\alpha}{h}. \tag{3.82}$$

Again, $\tilde{\sigma}_{x_1}$ and $\tilde{\sigma}_{x_2}$ can be obtained from Equations (3.78) and (3.79). The lengths l_{s_1} and l_{s_2} of sticking refer to the rearward and forward region. It is assumed in [34] that Equations (3.80) and (3.81),

$$l_{s_1} + l_{s_2} = \frac{(0.5 \sim 2)(h_1 + h_2)}{2}. \tag{3.83}$$

In Equations (3.80) and (3.81), $\tilde{q}_1$ and $\tilde{q}_2$ are the frictional force intensities in the rearward and forward regions on the borders of the brake and stick zone located at $x = \tilde{x}_1 = R \sin \tilde{\alpha}_1$ and $x = \tilde{x}_2 = R \sin \tilde{\alpha}_2$. Note that $\tilde{x}_1 - x_0 = l_{s_1}$ and $x_0 - \tilde{x}_2 = l_{s_2}$.

The contact pressure and the frictional force intensity are given in Figure 3.14 for the cases of strips made of material $m_1 = 1/7$; $m_2 = 0.2$ (solid lines) and $m_1 = 1/7$;

$m_2 = 0$ (dash lines). The values $h_1 = 11.6$ mm, $h_2 = 6.6$ mm, $R = 375$ mm, $\alpha_1 = 6.61^\circ$ and $f = 0.3$ are also used.

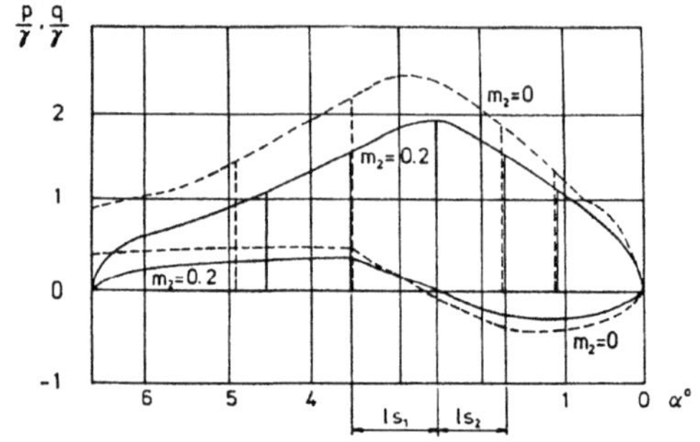

Fig. 3.14: Contact pressure and frictional force in strip rolling.

3.4 Two-dimensional problems

While the one-dimensional treatment does not give the details of the local stresses and strains, it does yield useful information on the strength. Two- and three-dimensional problems are more complicated; methods applying finite elements and variational principles may be used. For instance, the principle of minimum potential energy has been used [40, 41] to establish an upper bound solution for the extrusion of a nonlinear-viscous material through a conical die. The displacement finite element method has been used to solve the extrusion of a round rod through a plane die [42] and the deformation of solid and hollow circular cylinders [43, 44]. Membrane shell theory [45] was used to solve the large deformation of a circular membrane made of nonlinear viscous material.

Extrusion of rod

Consider the extrusion of a round rod of diameter d_1 and length l as shown in Figure 3.15. Both the punch with velocity v and the die with an opening diameter d_2 are rigid. Frictions are neglected and the solution to this problem can be found in [42].

If v_r and v_z are, respectively, the radial and axial velocities, then the boundary condition for $z = 0$ can be written as

$$v_z = 0;\ r \geq \frac{d_2}{2}$$

and for $z = l$ as

$$v_r = 0;\ r = \frac{d_1}{2} \quad \text{and} \quad v_z = -v;\ 0 \leq r \leq \frac{d_1}{2}.$$

Dividing the time deformation into small increments, the displacement and strain rates are related as

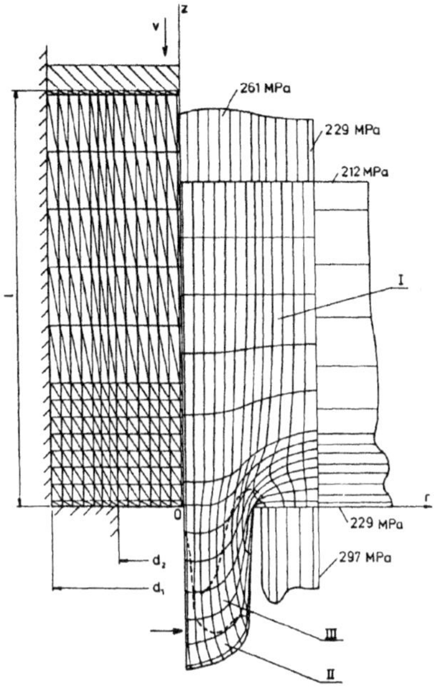

Fig. 3.15: Finite element grid pattern in undeformed (left side) and deformed (right side) state.

$$\{\xi\} = [B]\{\dot{q}\}, \tag{3.84}$$

where

$$\{\xi\} = \{\xi_z, \xi_r, \xi_t, \eta_{rz}\}^T$$

is the column vector of the strain rates, and ξ_z, ξ_r, ξ_t are the strain rates in the axial, radial, circumferential directions, respectively. The component η_{rz} is the angular strain and $\{\dot{q}\}$ is the column vector of the nodal velocities while $[B]$ is the matrix determined by the displacement velocities over a finite element volume. An axisymmetric ring with a triangular cross section is used as the finite element shape [46].

For $\nu = 0.48$, Equation (3.14) gives

$$\{\sigma\} = \frac{\sigma_e}{\xi_e}[\mathscr{D}]\{\xi\}, \tag{3.85}$$

where

$$\{\sigma\} = \{\sigma_z, \sigma_r, \sigma_t, \tau_{rz}\}^T$$

is the column vector for the stresses σ_z, σ_r and σ_t that are, respectively, in the axial radial and circumferential direction and τ_{rz} is the shear stress. The matrix $[\mathscr{D}]$ is [46]

$$[\mathscr{D}] = \frac{1-\nu}{(1+\nu)(1-2\nu)} \begin{vmatrix} 1 & \frac{\nu}{1-\nu} & \frac{\nu}{1-\nu} & 0 \\ 0 & 1 & \frac{\nu}{1-\nu} & 0 \\ 0 & 0 & 1 & 0 \\ 0 & 0 & 0 & \frac{1-2\nu}{2(1-\nu)} \end{vmatrix}. \tag{3.86}$$

The strain hardening expression in Equation (3.18) is applied in the step-by-step calculation to yield

$$\{\sigma\} = A\xi_e^{m_1-1}\left(\sum_{j=1}^{i} \xi_{ej}\Delta t\right)^{m_2} [\mathscr{D}]\{\xi\}, \tag{3.87}$$

where ξ_{ej} is the constant effective strain rate for the jth time step Δt.

The stiffness matrix is given by

$$[K] = 2\pi A\xi_e^{m_1-1}\left(\sum_{j=1}^{i} \xi_{ej}\Delta t\right)^{m_2} [\overline{B}]^T[\mathscr{D}][\overline{B}]r_c\Delta, \tag{3.88}$$

in which Δ and r_c are the area and center coordinate of the element. The equation governing the nodal velocity $\{\dot{\delta}\}$ is

$$[K]\{\dot{\delta}\} = \{R\}, \tag{3.89}$$

where $\{R\}$ is the nodal force column vector. The system of Equations (3.89) is integrated by iteration and the solution at each time step is obtained by the method of variable parameters. Accumulating the results of each time step gives the sum $\Sigma_{j=1}^{1} \xi_{ej}\Delta t$.

The extrusion of a round steel rod at 1150°C was calculated. The constants A, m_1 and m_2 for this material are the same as those for the example of a narrow rectangular membrane. The values of $d_1 = 120$ mm, $d_2 = 50$ mm, $l = 160$ mm and $v = 0.9$ mm/sec are used. A total of 600 finite elements is used. The undeformed and deformed grid pattern are on the left and right hand side in Figure 3.15, respectively. The punch displacements is $s = 0.216l$. Because of the absence of friction, the rod is deformed appreciably near the die exit and remains flat near the punch. Three different types of stress states are also shown in Figure 3.15. They are labelled as I for the nonuniform compression, II for the three-dimensional nonuniform tension and III for the three-dimensional stress state with different signs of maximum and minimum principal stresses.

Referring to region II in Figure 3.15, the time history of σ_r and σ_t (curve 1) and σ_y (curve 2) are shown graphically in Figure 3.16. Note that a sign change occurred at approximately $t = 14$ sec where compression changes to tension or vice versa. Figure 3.17 shows the time dependence of F, the extrusion load.

Deformation of solid and hollow circular cylinder

The method described in the previous section also applies to the deformation of solid

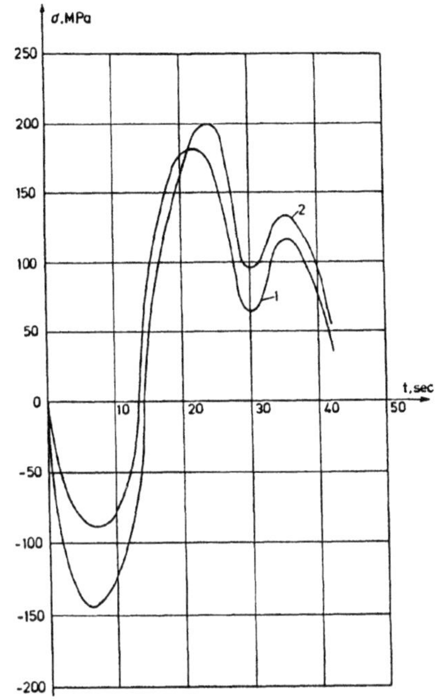

Fig. 3.16: Time history of σ_r and σ_t (curve 1) and σ_y (curve 2) in region II of Figure 3.15.

and hollow cylinders. Equations (3.87) and (3.88) are changed to

$$\{\sigma\} = (\xi_e^{1-1/n} B^{1/n}(t))^{-1}[\mathscr{D}]\{\xi\}, \tag{3.90}$$

and

$$[K] = 2\pi(\xi_e^{1-1/n} B^{1/n}(t))^{-1}[\overline{B}]^T[\mathscr{D}][\overline{B}]r_c\Delta. \tag{3.91}$$

The boundary conditions are $v_z = 0$ for $z = 0$ and $v_z = -v/2$; $v_r = 0$ for $z = h$.

Results are obtained for the deformation of a heated aluminum solid cylinder whose diameter is equal to its length. With $n = 5$ in Equation (3.17), the variations of $b(t)$ with time are given in Figure 3.18. The relative velocities of the plates are constant; they are 0.167 mm/sec and $v = 0.667$ mm/sec. Experimental data on specimens with $h_0 = D_0/2 = 12.5$ mm can be found in [47]. Figure 3.19 shows the deformed pattern for Poisson ratio $\nu = 0.44$ (curve 1) and $\nu = 0.48$ (curve 2). A total of 512 elements were used. Curve 3 corresponds to the experimental data. The accuracy of the solution increases with ν and computer time. The results shown in Figure 3.20 are for two velocities of deformation. Type II discretization using only 384 elements is more effective than type I which used 512 elements. Refer to [44] for similar results in the case of a hollow cyliner.

3.5 Damage in metal forming

The theory in [48] can be used to evaluate the development of cracks in hot metal

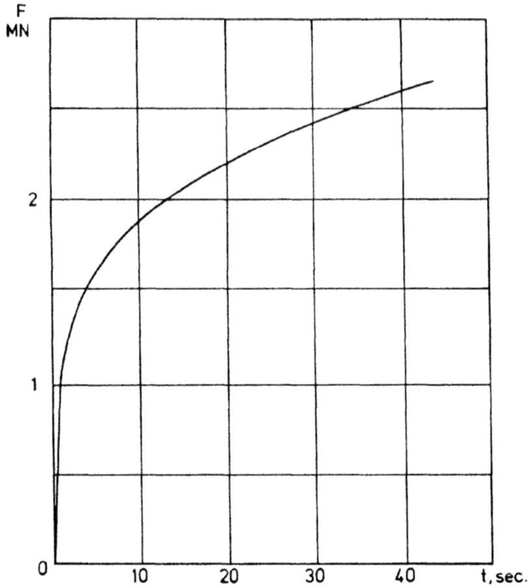

Fig. 3.17: Time dependence of extrusion load.

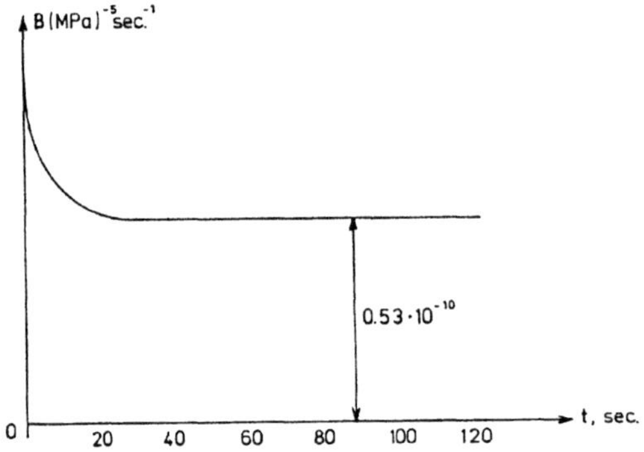

Fig. 3.18: Time dependence of $B(t)$.

forming. The material is initially continuous with $\psi = 1$ while final rupture corresponds to $\psi = 0$. The equation that describes material continuity is given by

$$\frac{d\psi}{dt} = -A\left(\frac{\sigma_{er}}{\psi}\right)^m \tag{3.92}$$

where A and m are material constants at a given temperature and σ_{er} is an equivalent stress for rupture [8]. Rupture time can be obtained from the condition that $\sigma_{er} = \sigma_{max}$.

Equation (3.92) may be integrated to give

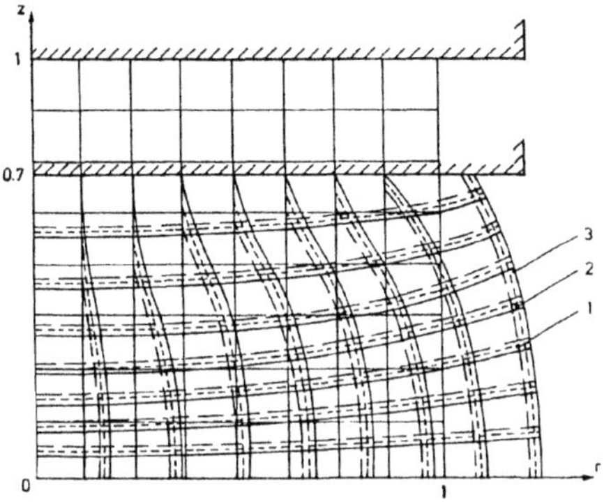

Fig. 3.19: Deformed grid for ν = 0.44 (curve 1), ν = 0.48 (curve 2) and experimental (curve 3).

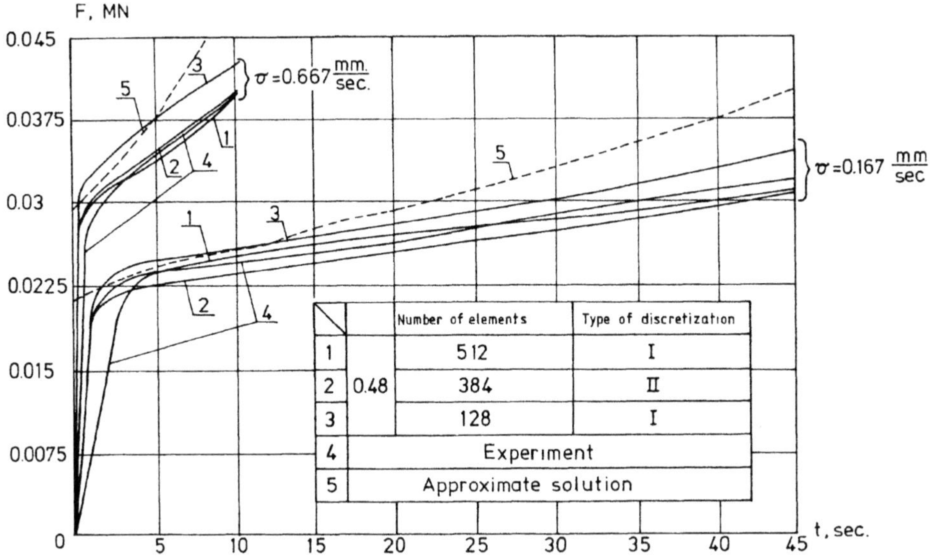

Fig. 3.20: Time history of deformation load: theory and experiment.

$$\psi^{m+1} = 1 - \frac{A}{m+1} \int_0^t \sigma_{er}^m \, dt. \tag{3.93}$$

The case $\psi = 0$ gives the inhibited time of rupture where a rupture front is developed. As time increases, the material can be fully ruptured, a condition that is not of interest in metal forming. Equation (3.93) correlates damage with continuity while creep is not taken into account. It was suggested in [6] that

$$\xi = \frac{k_1 \sigma^n}{(1-\omega)^{b_1}}, \quad \frac{d\omega}{dt} = \frac{k_2 \sigma^m}{(1-\omega)^{b_2}}, \tag{3.94}$$

where $\omega = 1 - \psi$. Damage and creep can thus be related [49 – 53]. If $m = n$ and $b_1 = b_2$, then Equation (3.94) gives

$$\omega = \frac{\overline{\varepsilon}}{\overline{\varepsilon_*}}$$

where $\overline{\varepsilon_*}$ is the logarithmic strain at rupture. The equation for creep can be obtained as:

$$t = \frac{1}{\xi_0} \int_1^{\varsigma} \varsigma^{-n-1} (1-\omega)^{b_1} \, d\varsigma, \tag{3.95}$$

where $\xi_0 = k_1 \sigma_0^n$ is the initial strain rate for the stress σ_0 and $\varsigma = 1 + \varepsilon$.

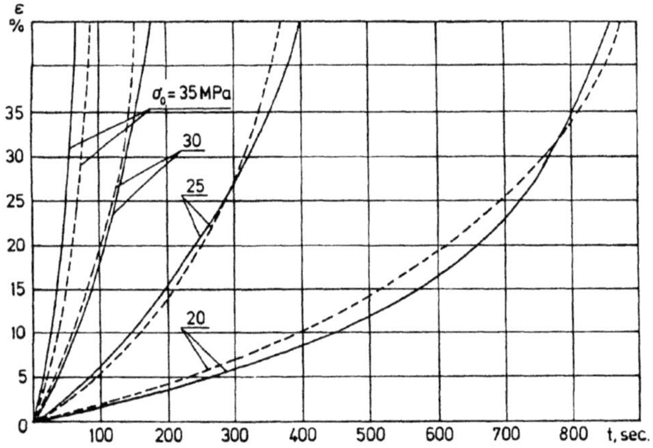

Fig. 3.21: Creep curves for aluminum alloy at 420°: experiment (solid) and calculation (dotted).

Experimental creep curves for an aluminum alloy at 420° are shown by the solid lines in Figure 3.21 [54]. Equation (3.95) has been used to calculate the dotted lines using $n = 3.97$, $k_1 = 138 \times 10^{-9}$ (MPa)$^{-n}$ sec^{-1}, $k_2 = 235 \times 10^{-y}$ (MPa)$^{-n}$ sec^{-1}, $b_1 = 0.4$. The values of ς^* at $\sigma_0 = 20, 25, 30, 35$ MPa are equal to 1.74, 1.76, 1.91 and 1.79, respectively.

The equations governing creep rupture of specimens at different lengths are [52, 53]

$$\xi = \frac{k \sigma^n}{(1-\omega)^{b_1}}, \quad \frac{d\omega}{dt} = \frac{\sigma \xi}{H}. \tag{3.96}$$

It follows from Equation (3.96) that $\omega = \sigma_0 \varepsilon / H$ and $\varepsilon_* = H / \sigma_0$. Therefore,

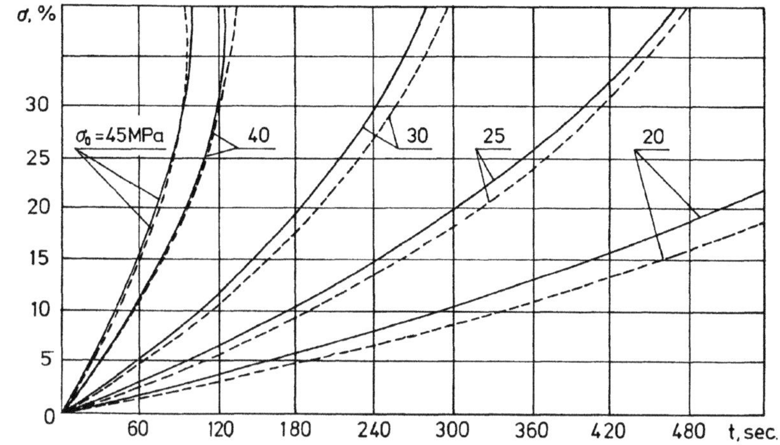

Fig. 3.22: Creep curves for magnesium alloy at 380°: experiment (solid) and calculation (dotted).

$$t = \frac{1}{k\sigma_0^n}\int_0^{\varepsilon}\left(\frac{1-\sigma_0\varepsilon}{H}\right)^{b_1}(1+\varepsilon)^{-n-1}\,d\varepsilon. \tag{3.97}$$

Experimental creep curves for the magnesium alloy at 380°C are presented in Figure 3.22 by the solid lines. The dotted curves are computed from Equation (3.97) with $n = 2.5$, $b_1 = 1.12$, $H = 40$ MPa, $K_1 = 1.37\times10^{-1}$ $(\mathrm{MPa})^{-n}$ sec^{-1} [54]. Damage can also be associated with strain localization which can lead to fracture as has been observed experimentally. Such evaluation in cold forming has been made by application of Drucker's postulate, which can be generalized to account for time dependent material properties [55, 56]. The works in [57 – 59] are also relevant to this problem.

References

[1] HENCKY, H., *Zeitschr. Angew. Math. und Mech.*, Vol. 5, 1925, p. 115.

[2] ISHLINSKI, A.Yu., *Prikladnaja Matematika i Mehanika*, Vol. VII, 1943, p. 226.

[3] TARNOVSKI, I.Ja., POZDEEV, A.A., BAAKASHVILI, V.S., MEANDROV, L.V., TARNOVSKI, V.I. and HASIN, G.A., *Resistance to Deformation and Plasticity of Steel at High Temperatures*, Sabchota Sakartvelo, Tbilisi, 1970, p. 173.

[4] POZDEEV, A.A., TARNOVSKI, V.I., EREMEEV, V.I. and BAAKASHVILI, V.S., *Application of Creep Theory to Metal Forming*, Metallurgia, Moscow, 1973, p. 39.

[5] ILJUKOVICH, B.M., BAAKASHVILI, V.S. and BEDINEJSHVILI, R.V., *Theoretical Bases of Plastic Metal Working*, Sabchota Sakartvelo, Tbilisi, 1979, p. 560.

[6] RABOTNOV, Yu.N., *Creep of Structure Elements*, Nauka, Moscow, 1966, p. 212, 223.

[7] MALININ, N.N., *Applied Theory of Plasticity and Creep*, Mashinostroenie, Moscow, 1975, p. 267.

[8] MALININ, N.N., *Creep Calculations of Mechanical Engineering Structure Elements*, Moscow, 1981, p. 24.

[9] DAVENPORT, C.C., *J. Appl. Mech.*, Vol. 5, 1938, p. A-55.
[10] KACHANOV, L.M., *Creep Theory*, Fitzmagiz, Moscow, 1960, p. 39.
[11] LUDVIK, P., *Elements of Technological Mechanics*, Verlag von J. Springer, 1909, p. 44.
[12] NADAI, A., The Influence of Time upon Creep, The Hypberbolic Sine Creep Law, in *Stephen Timoshenko Anniversary Volume*, D.H. Young, ed., Macmillan, London, 1938, p. 165.
[13] SHESTERIKOV, S.A. *Izvestija A.N. SSSR Otdelenie Tehnickeshik Nauk.*, Vol. 2, 1957, p. 122.
[14] GRABSKI, M.W., *Structural Superplasticity of Metals*, Slask. Katowice, 1973, p. 22.
[15] SMIRNOV, O.M., *Plastic Metal Working in Superplastic Condition*, Mashinostroenie, Moscow, 1979, Vol. 17, p. 126.
[16] ZIUZIN, V.I., BROVMAN, M.Ja and MELNIKOV, A.F., *Resistance to Deformation of Steels at Hot Rolling*, Metallurgija, Moscow, 1964, p. 211.
[17] TRETJAKOV, A.V. and ZIUZIN, V.I., *Mechanical Properties of Metals and Alloys at Plastic Working*, Metallurgija, Moscow, 1973, p. 25.
[18] RABOTNOV, Yu.N. and MILEIKO, S.T., *Short-Time Creep*, Nauka, Moscow, 1970, p. 32.
[19] MALININ, N.N., *Arch. Mech.*, Vol. 24, 1972, p. 439.
[20] MALININ, N.N., Creep in Metal Working, in *Teoretichna i Prilozna Mehanika*, Kniga 2, Vtori Kongress, Varna, G. Brankov, ed., Izdatelstvo na Blgarskata Akademija na Naukite, Sofia, 1976, p. 143.
[21] MALININ, N.N., *Mechanics of Creep in Metal Forming*, IUTAM Symposium, Tutzin F.R.G., (1978), H. Lippmann, ed., Springer-Verlag, Berlin, 1979, p. 318.
[22] MALININ, N.N., Creep in Metal Working, in *Teoretichna i Prilozna Mehanika.* Kniga I. Chetvrti Kongress, Varna (1981), G. Brankov, ed., Izdatelstvo na Blgarskate Akademija na Naukite, Sofia, 1981, p. 437.
[23] MALININ, N.N., *Izvestija Vuzov. Mashinostroenie*, Vol. 5, 1982, p. 112.
[24] ZIENKIEWICZ, O.C. and GODBOLE, P.N., *Int. J. Num. Meth. Eng.*, Vol. 8, 1974, p. 3.
[25] CRISTESCU, N., *Int. J. Mech. Sci.*, Vol. 17, 1975, p. 125.
[26] SEGAL, V.M., *Technological Problems of Theory of Plasticity*, Nauka i Tehnika, Minsk, 1977, p. 207.
[27] GUN, G.Ja., *Theoretical Bases of Plastic Metal Working, Theory of Plasticity*, Metallurgija, Moscow, 1980, p. 373.
[28] MALININ, N.N. and SHIRSHOV, A.A., *Izvestija Vuzov. Mashinostroenie*, Vol. 2, 1978, p.145.
[29] MALININ, N.N. and ROMANOV, K.I., Stability of Material in Hot Forming, in *Proceedings of 4th International Conference on Production Engineering*, Tokyo (1980), H. Kudo, ed., Jap. Soc. Tech. Plast., 1981, p. 172.
[30] MALININ, N.N. and ROMANOV, K.I., *Mashinovedenie*, Vol. 4, 1982, p. 98.
[31] MALININ, N.N., Deformation of Heated Thin-Walled Tubes, in *Raschety na Prochnost.* Vyp. 19, N.D. Tarabasov, ed., Mashinostroenie, Moscow, 1982, p. 102.
[32] HOLT, D.L., *Int. J. Mech. Sci.*, Vol. 12, 1970, p. 491.
[33] STOROZEV, M.V. and POPOV, E.A., *Theory of Plastic Metal Working*, Mashinostroenie, Moscow, 1977, p. 321.
[34] TSELIKOV, A.I., NIKITIN, G.S. and ROKOTYAN, S.E., *Theory of Lengthwise Rolling*, Metallurgija, Moscow, 1980, p. 46.
[35] JOHNSON, W. and MELLOR, P.B., *Engineering Plasticity*, Van Nostrand Reinhold Company, London, 1973, pp. 338, 350.
[36] MALININ, N.N., *Izvestija Vuzov. Mashinostroenie*, Vol. 12, 1977, p. 119.
[37] MALININ, N.N. and ROMANOV, K.I., *Izvestija Vuzov. Mashinostroenie*, Vol. 7, 1977, p. 104.

[38] TSELKOV, A.I., *Rolling Mills,* Metallurgizdat, Moscow, 1946, p. 560.
[39] UNKSOV, E.P., *Engineering Theory of Plasticity,* Mashigz, Moscow, 1959, p. 143.
[40] BOBROVNIKOVA, N.N., *Izvestija Vuzov. Mashinostroenie,* Vol. 5, 1973, p. 130.
[41] BOBROVNIKOVA, N.N., *Izvestija Vuzov. Mashinostroenie,* Vol. 1, 1974, p. 133.
[42] ROMANOV, K.I. and KALMYKOVA, N.V., *Izvestija Vuzov. Mashinostroenie,* Vol. 9, 1982, p. 9.
[43] MALININ, N.N. and ROMANOV, K.I., *Izvestija Vuzov. Mashinostroenie,* Vol. 8, 1977, p. 127.
[44] ROMANOV, K.I., *Mashinovedenie,* Vol. 5, 1978, p. 79.
[45] GAVRJUSHINA, N.T., *Izvestija Vuzov. Mashinostroenie,* Vol. 3, 1982, p. 29.
[46] ZIENKIEWICZ, O.C., *The Finite Element Method in Engineering Science,* McGraw-Hill, London, 1971, p. 16.
[47] ROMANOV, K.I., *Izvestija Vuzov. Mashinostroenie,* Vol. 6, 1977, p. 147.
[48] KACHANOV, L.M., *Bases of Fracture Mechanics,* Nauka, Moscow, 1974, p. 138.
[49] MELNINKOV, G.P. and SHESTERIKOV, S.A., *Zurnal Prikladnoi Mehaniki i Technicheskoi Fiziki,* Vol. 2, 1972, p. 91.
[50] LOKOSCHENKO, A.M. and SHESTERIKOV, S.A., *Zurnal Prikladnoi Mehaniki i Technicheskoi Fiziki,* Vol. 3, 1980, p. 155.
[51] LOKOSCHENKO, A.M. and SHESTERIKOV, S.A., *Zurnal Prikladnoi Mehaniki i Technicheskoi Fiziki,* Vol. 1, 1982, p. 160.
[52] SOSNIN, O.V. and TORSHENOV, M.G., *Problemy Prochnosti,* Vol. 7, 1972, p. 55.
[53] SOSNIN, O.V., GOREV, B.V. and NIKITENKO, A.F., *Problemy Prochenosti,* Vol. 11, 1976, p. 3.
[54] LAZARENKO, E.S., MALININ, N.N. and ROMANOV, K.I., *Izvestija Vuzov. Machinostroenie,* Vol. 3, p. 25, Vol. 7, p. 19.
[55] DRUCKER, D., *J. Appl. Mech.,* Vol. 26, 1959, p. 101.
[56] DRUCKER, D., *J. de Mec.,* Vol. 3, 1964, p. 235.
[57] MALININ, N.N. and ROMANOV, K.I., Stability of Tension of a Bar in Creep Condition, in *Raschety na Prochnost,* Vyp. 21, D.N. Tarabasov, ed., Mashinostroenie, Moscow, 1980, p. 104.
[58] MALININ, N.N. and ROMANOV, K.I., *Izvestija A.N. SSSR Mehanika Twerdogo Tela,* Vol. 1, 1981, p. 133.
[59] ROMANOV, K.I., *Mashinovedenie,* Vol. 5, 1980, p. 75.

L.V. Nikitin and Sh.A. Mukhamediev

4

Failure of inelastic solids

4.1 Introduction

Solids can fail mechanically under the action of external forces. Two extreme types of failure behavior are brittle fracture, when a solid breaks without substantial permanent deformation, and viscous rupture, where at least one of the dimensions of the solid is very small. The former is frequently associated with a tensile mode and latter with a shear mode with plastic deformation. In practice, failure may contain both of the foregoing types – i.e., mixed failure mode – and the behavior applies to a specimen or structure.

A structure can lose its load carrying capacity if it is deformed excessively without load increase before or at the same time as complete failure. Instability of geometric or rheological nature, unstable damage accumulation or subcritical crack growth can precede the full loss of structural integrity; these events can also occur simultaneously. The thresholds of global failure can be referred to as the critical states that involve various types of instability and/or loss of load bearing capacity and/or crack growth. Deformation in the subcritical states may be elastic or plastic, depending on the material properties. Microfracture and/or stable damage accumulation should be considered on a phenomenological level in the subcritical states. The sequence of events is shown schematically in Figure 4.1(a). The choice of a given path depends on the geometry, constitutive relation and load condition. It is also possible for the process to terminate at some post-critical state. By determining if the stress and strain state in an element is homogeneous, then the loss of load carrying capacity can be of the ideal viscous rupture type as indicated in Figure 4.1(b); this should be distinguished from that in a structure where the stress and strain can be inhomogeneous.

The present paper addresses some of the more recently investigated problems of failure in solids that undergo inelastic deformation. Loss of loading capacity of structures made of an ideally elastic-plastic compressible material under plane strain is first considered. Elastic compressibility is accounted for. Full plastic structures do not imply the loss of the load carrying capacity. The Mode I crack in an ideal elastic-plastic material is also analyzed. For an elastic compressible material, the crack tip is completely embraced by the plastic zones. Similar results are obtained from the deformation and incremental theory of plasticity. Crack growth in a linear viscoelastic solid is also studied without the cohesive force zone. Finally, rheological instability for structures made of elastic-plastic materials is considered by including the effects of internal friction and volume change. Constraints of boundary conditions on localized plastic deformation are discussed in connection with the integral criterion.

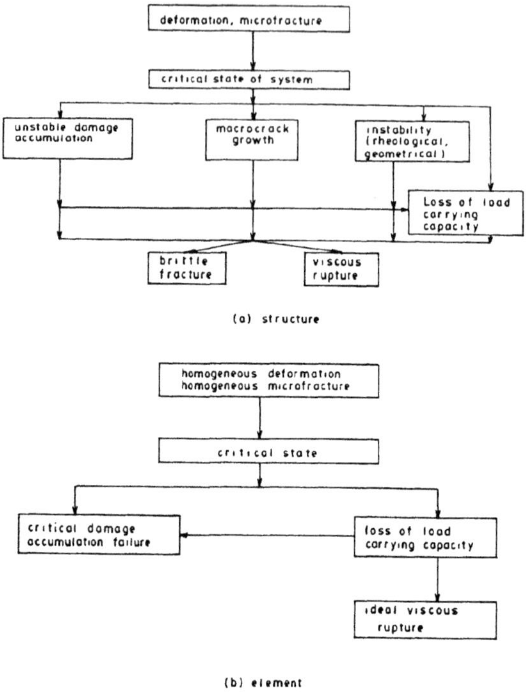

Fig. 4.1: Failure sequence: (a) structure and (b) element.

4.2 Load bearing capacity of an ideal elastic-plastic solid under plane strain

The load bearing capacity of structures is found for the rigid plastic model [1, 2]. The available solutions of the elastic-plastic problems [2] show that elasticity affect the results. For the sake of simplicity, the elastic compressibility is often neglected; it does, however, affect the displacements prior to the loss of load-carrying capacity. The inclusion of compressibility in the elastic region only is not adequate as it can lead to discontinuous solutions. An ideally plastic elasto-compressible material [3] will therefore be used.

Equations of plane strain

The deformation plasticity theory constitituve relation for an elasto-compressible material in the terms of the principal stresses σ_α and strains ε_α take the form

$$\varepsilon_\alpha = K\sigma + \psi(\sigma_\alpha - \sigma) \tag{4.1}$$

where $\sigma = 1/3(\sigma_1 + \sigma_2 + \sigma_3)$ is the mean stress and K is the bulk modulus. The convention of summation will be adopted; the Greek indices are to be summed over 1, 2, 3, but the Latin ones over 1, 2.

Consider an elastic-plastic material which is nonhardening in shear and is described by the Prandtl diagram. The conditions of the ideal plasticity are assumed either in the forms of the Mises condition

$$\tau = \sqrt{\frac{1}{6}}((\sigma_1 - \sigma_2)^2 + (\sigma_2 - \sigma_3)^2 + (\sigma_3 - \sigma_1)^2)^{1/2} = \tau_S \tag{4.2a}$$

or the Tresca condition

$$\max\left\{\frac{|\sigma_1 - \sigma_2|}{2}, \frac{|\sigma_2 - \sigma_3|}{2}, \frac{|\sigma_3 - \sigma_1|}{2}\right\} = \tau_S \tag{4.2b}$$

or the condition

$$\max(|\sigma_1 - \sigma|, |\sigma_2 - \sigma|, |\sigma_3 - \sigma|) = \tau_S \tag{4.2c}$$

where τ_S is the yield stress in shear.

Equations (4.1), (4.2) do not permit an explicit expression of Ψ in terms of the stresses. It is therefore not possible to determine the strain components uniquely in terms of the stress in the plastic state. The condition of plane strain $\varepsilon_3 = 0$ simplifies the situation. For the incompressible material $K = 0$ so that

$$\sigma_3 = p, \ p = \frac{1}{2}(\sigma_1 + \sigma_2) \tag{4.3}$$

and all three conditions of plasticity take the same form

$$s = \tau_S, \ s = \frac{1}{2}|\sigma_1 - \sigma_2|. \tag{4.4}$$

For a compressible material one has as a result:

$$\sigma_3 = 2\chi p, \ \chi = \frac{\Psi - K}{2\Psi + K}$$

$$\frac{1}{2\mu} \leq \Psi < \infty, \ \nu \leq \chi < \frac{1}{2} \tag{4.5}$$

$$\tau = \left\{s^2 + \frac{1}{3}(1 - 2\chi)^2 p^2\right\}^{1/2}$$

where μ is the rigidity and ν is Poisson's ratio. Any of the conditions in Equations (4.2) using (4.5) may be presented as the plasticity curve in the principal stress space σ_1, σ_2:

$$f(\sigma_i, \chi) = 0.$$

The coefficient χ, being a generalization of Poisson's ratio, characterizes the plastic deformation accumulation and varies from ν in the elastic region to 1/2 for unbounded plastic deformation. With increasing χ, the plastic curve $f = 0$ approaches the plastic curve in Equation (4.4) for an incompressible material. The coefficient χ plays the role of strain hardening parameter and incorporates compressibility under plane strain giving rise to hardening behavior for a nonhardening material.

For the plane strain $\varepsilon_3 = 0$, from Equation (4.1), the in-plane strains are

$$\varepsilon_i = \Psi(\sigma_i - 2\chi P). \tag{4.6}$$

Using the yield condition in Equations (4.2), it is found for Ψ for the Mises condition

$$\Psi = \frac{K}{2}\left(\frac{\sqrt{3}\,|P|}{\sqrt{\tau_S^2 - S^2}} - 1\right) \tag{4.7a}$$

for the Tresca condition

$$\Psi = \frac{K}{2}\left(\frac{3\,|P|}{2\tau_S - S} - 1\right) \quad \text{for } S < \tau_S \tag{4.7b}$$

and for the condition in Equations (4.2c)

$$\Psi = \frac{K}{2}\left(\frac{2\,|P|}{\tau_S} - 1\right) \quad \text{for } 2S \le \tau_S$$

$$\Psi = \frac{K}{2}\left(\frac{|P|}{\tau_S - S} - 1\right) \quad \text{for } \tau_S \le 2S < 2\tau_S. \tag{4.7c}$$

When $S = \tau_S$ in Ψ, the Tresca condition cannot be expressed uniquely through the stresses as in the case of an incompressible material. For the Prandtl–Reuss incremental theory of plasticity, the principal strain increments have the form

$$\mathrm{d}\,\varepsilon_\alpha = \mathrm{d}\,\varepsilon_\alpha^e + \mathrm{d}\,\lambda(\sigma_\alpha - \sigma) \tag{4.8}$$

where $\mathrm{d}\,\varepsilon_\alpha^e$ are the elastic deformation increments and the second term in Equation (4.8) represents the plastic deformation increments.

Consider the Mises condition of plasticity in Equation (4.2a). From Equation (4.8) and the planar condition $\mathrm{d}\varepsilon_3 = 0$, it is easy to obtain the differential equation with respect to σ_3. The solution of this equation with continuity $\sigma_3 = 2\nu p$ preserved across the elastic-plastic boundary has the form

$$\sigma_3 = 2\nu p + \frac{2}{3}\frac{(1-2\nu)^2}{K}\exp\left(-\frac{2}{3}\frac{(1-2\nu)\lambda}{K}\right) \times$$

$$\times \int_0^\lambda P \exp\left(\frac{2}{3}\frac{(1-2\nu)\lambda^1}{K}\right) \mathrm{d}\lambda^1. \tag{4.9}$$

From Equation (4.9) and the Mises condition in Equation (4.2a), it follows that

$$\mathrm{d}\lambda = \frac{\mathrm{d}((1-2\nu)P - Q)}{\frac{2}{3}\frac{1-2\nu}{K}Q}$$

$$Q = \mathrm{sign}(P)\sqrt{3(\tau_S^2 - S^2)} \tag{4.10}$$

In view of Equations (4.10) and (4.8), the theory of plasticity for a compressible ideal elastic-plastic material can uniquely express the strain increments in terms of the in-plane stresses and their increments. From Equation (4.6), the deformations ε_r, ε_ϕ, $\phi_{r\phi}$ in the polar system of coordinates r, ϕ are given by

$$\varepsilon_r = K\left(\frac{3}{2} - v\right)\left(P + \frac{S}{2v}\right)$$

$$\varepsilon_\phi = K\left(\frac{3}{2} - v\right)\left(P - \frac{S}{2v}\right) \tag{4.11}$$

$$\varepsilon_{2\phi} = K\left(\frac{3}{2} - v\right)\frac{\sigma_{2\phi}}{2v}$$

where $p = 1/2(\sigma_r + \sigma_\phi)$, $S = 1/2(\sigma_r - \sigma_\phi)$, $v = 1/2 - \chi$, and σ_r, σ_ϕ, $\sigma_{r\phi}$ are the stress components. The complete system of the equations for plane strain of a compressible ideally elastic-plastic material with the Mises yield condition in terms of stresses consists of the equilibrium equations

$$\frac{\partial (P + S)}{\partial r} + \frac{2S}{r} + \frac{1}{r}\frac{\partial \sigma_{r\phi}}{\partial \phi} = 0$$

$$\frac{\partial (P - S)}{r\partial \phi} + \frac{\partial \sigma_{r\phi}}{\partial r} + \frac{2\sigma_{r\phi}}{r} = 0 \tag{4.12}$$

the compatibility equation

$$v_3\left(\frac{3}{4} - v + v^2\right)r^2\Delta P - P\,v^3 r^2 \Delta v$$

$$- v\left(\frac{3}{2} + 2v^2\right)\left(r^2\frac{\partial P}{\partial r} + \frac{\partial v}{\partial r} + \frac{\partial P}{\partial \phi}\frac{\partial v}{\partial \phi}\right)$$

$$+ \frac{3}{4}Sv\left(r^2\frac{\partial^2 v}{\partial r^2} - r\frac{\partial v}{\partial r} - \frac{\partial^2 v}{\partial \phi^2}\right)$$

$$+ \frac{3}{2}S\left(\left(\frac{\partial v}{\partial \phi}\right)^2 - r^2\left(\frac{\partial v}{\partial r}\right)^2\right)$$

$$+ \frac{3}{2}\sigma_{r\phi}v\left(r\frac{\partial^2 v}{\partial r \partial \phi} - \frac{\partial v}{\partial \phi}\right) - 3\sigma_{r\phi}r\frac{\partial v}{\partial r}\frac{\partial v}{\partial \phi} = 0 \tag{4.13}$$

$$\Delta = \frac{\partial^2}{\partial r^2} + \frac{1}{r}\frac{\partial}{\partial r} + \frac{1}{r^2}\frac{\partial^2}{\partial \phi^2}$$

and the Mises yield condition

$$S^2 + \sigma_{r\phi}^2 + \frac{4}{3}v^2 p^2 \leq \tau_S^2. \tag{4.14}$$

They are of the elliptic type. The only exception is the case of the pure shear $p = 0$ and $S = \tau_S$ for the Tresca condition. In these cases, a second order system of hyperbolic type is obtained. For an elastically compressible material under plane strain, boundary value problems in terms of the stresses may be formulated for purely plastic states.

Solution to Equations (4.12), (4.13) and (4.14) for an incompressible material corresponds to small values of the parameter $\delta = 1/2 - \nu$. The function p, S, $\sigma_{r\phi}$, and v can be expanded in asymptotic series in terms of δ:

$$F(r, \phi, \delta) = \sum_{i=0} \delta^i F_i(r, \phi), \; i = 1, 2, 3, \cdots \tag{4.15}$$

where F corresponds to one of the above functions. The solution for an incompressible material is the undisturbed state or zero approximation. Successive approximations can be applied in the conventional manner [4]. The conditions of coupling of solution are met at the undisturbed elastic-plastic boundary.

Viscous rupture of material element

The loss of load carrying capacity for a rigid plastic body can be obtained as a limiting case without analyzing the growth of plastic zone. For an ideal incompressible elastic-plastic material, the loss of load carrying capacity corresponds to the onset of unbounded plastic flow in view of the absence of a unique solution for the corresponding limit load. The case of a compressible material is different. Full plasticity is not sufficient for unbounded plastic deformations to take place. The load which corresponds to the onset of full plasticity does not necessarily coincide with the limit load.

To achieve the goal of Section 4.2, the stresses σ_i are scaled by τ_S, strains are scaled by $\tau_S/2\mu$; that is $K = (1 - 2\nu)/(1 + \nu)$. The element of an ideally elastic compressible plastic material is subjected to biaxial stresses σ_1 and $\sigma_2 = m\sigma_1$, where $\sigma_1 > 0$, and $0 \leq m < 1$. The results $p = \sigma_1(1 + m)/2$ and $s = \sigma_1(1 - m)/2$ then follow. It is assumed that m remains constant during deformation. For an incompressible material, the loss of load carrying capacity takes place at the onset of plasticity, i.e. $\sigma_1 = \overset{*}{\sigma}_{1(n)} = 2/(1 - m)$.

The plasticity state of a compressible element under the Mises yield condition in Equation (4.2a) takes place when σ_1 equals to $\overset{*}{\sigma}_1$:

$$\overset{*}{\sigma}_1 = 2\left\{(1 - m)^2 + \frac{1}{3}(1 + m)^2(1 - 2\nu)^2\right\}^{-1/2} \tag{4.16}$$

where $\overset{*}{\sigma}_1 \leq \sigma_1 < \overset{*}{\sigma}_{1(n)}$. Making use of Equations (4.6) and (4.2a), the result is

$$\varepsilon_1 = \frac{K}{2}\left(\frac{\sqrt{3}P}{\sqrt{1 - S^2}} - 1\right)\left(S + \sqrt{3(1 - S^2)}\right). \tag{4.17}$$

The loss in load carrying capacity for the compressible element ($\varepsilon_1 \to \infty$) occurs at the same stress $\sigma_1 = \overset{*}{\sigma}_{1(n)}$ as that for the incompressible element. Figure 4.2 shows the variations of σ_1 with ε_1 for the compressible material with the Poisson ratios ν_0 and $\nu(\nu_0 > \nu)$. Results for the other yield conditions in Equations (4.2) can be obtained in the same way. The dependence of σ_1 on ε_1 is analogous to that of the strain hardening material with the horizontal asymptotic line $\sigma_1 = \overset{*}{\sigma}_{1(n)}$.

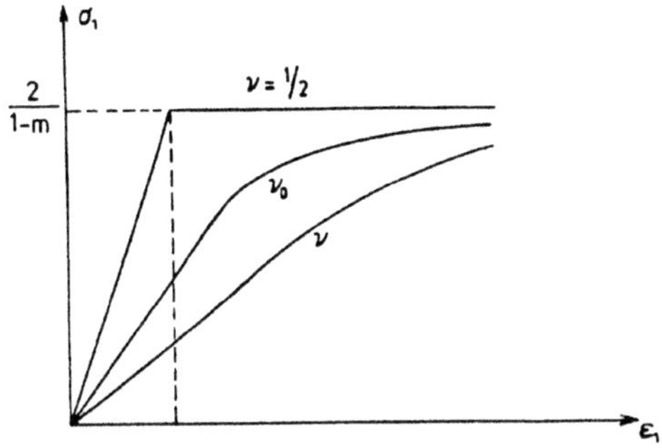

Fig. 4.2: Stress versus strain for an incompressible material ($\nu = 1/2$) and two compressible materials ($\nu_0 > \nu$).

Bending of a beam

Consider the elastic-plastic bending of a beam under plane strain in the plane x_1 and x_2, where x_1 is directed along the middle line of the beam. In this case, $\sigma_{22} = \sigma_{12} = 0$, $\varepsilon_{11} = \kappa x_2$ and $\kappa = -\mathrm{d}^2u_2/\mathrm{d}x_1^2$, where u_2 is the displacement along x_2 and σ_{ij} is the stress tensor. All length dimension values will be scaled by beam half-width h and the curvature κ will be scaled by $\tau_S/2\mu h$. The yield condition of the maximum stress in Equation (4.2c) is assumed. It can be shown that $2\sigma_1 - \sigma_3 = 3$ in the plastic zone. In the elastic zone, $\sigma_{11} = \sigma_1 = \kappa x_2/(1-\nu)$. Plasticity first appears at $x_2 = \pm 1$ when $|\kappa|^* = 3(1-\nu)/(2-\nu)$. For the curvature κ depends on the half-width of the elastic zone $\xi = 3(1-\nu)/|\kappa|(2-\nu)$. For $|\kappa| > |\kappa|^*$, the stress σ_1 is given by

$$\sigma_1 = \frac{\kappa x_2}{1-\nu}, \quad \text{for} \quad |x_2| \le \xi$$

$$\sigma_1 = \frac{\kappa x_2}{2K} + \mathrm{sign}(\kappa x_2) \times \left(2 - \sqrt{1 + (\kappa x_2/2K)^2}\right), \quad \text{for} \quad 1 > |x_2| > \xi.$$

The bending moment M per unit thickness in the elastic state is

$$M = \int_{-1}^{1} \sigma_1 x_2 \, \mathrm{d}x_2 = \frac{2\kappa}{3(1-\nu)}. \tag{4.18}$$

For $|\kappa| > |\kappa|^*$, the expression for $M(\kappa)$ has a form

$$M = \frac{\kappa}{3K}\left(1 - \left(1 + \frac{4K^2}{\kappa^2}\right)^{3/2}\right) +$$

$$+ 2\,\mathrm{sign}(\kappa)\left(1 + \frac{11 - 82\nu + 138\nu^2 - 82\nu^3 + 11\nu^4}{3\kappa^3(2-\nu)^2(1+\nu)^2}\right). \tag{4.19}$$

When $\nu = 1/2$, Equation (4.19) becomes analogous to the one for an incompressible material. The dependence of M on κ for the incompressible beam and for two compressible beams with Poisson's ratios ν_0 and ν, ($\nu_0 > \nu$) is shown in Figure 4.3.

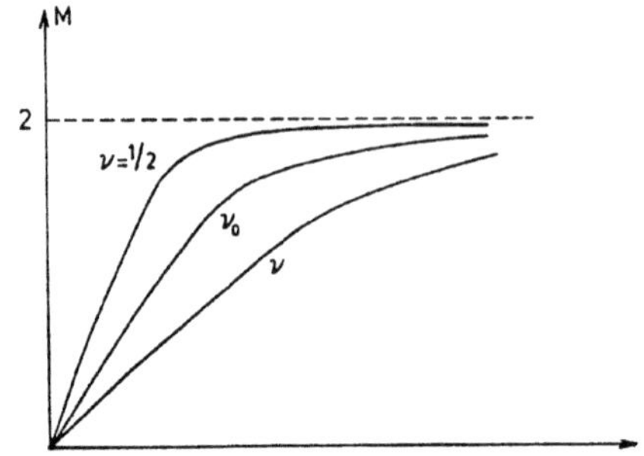

Fig. 4.3: Bending moment M versus curvature κ for an incompressible material ($\nu = 1/2$) and two compressible materials ($\nu_0 > \nu$).

Cylindrical tube under internal pressure

Solutions to the elastic-plastic problems of cylindrical tubes can be found in [2] which also contains the numerical evaluation of an ideal compressible elastic-plastic tube in plane strain.

The method of perturbation will be used. All quantities with dimension of length will be normalized in terms of the external tube radius b. The stress-strain state is assumed to depend on r only. The Mises yield condition is adopted. The boundary conditions are

$$(P + S)\,|_{r=a} = -q, \quad (P + S)\,|_{r=1} = 0 \tag{4.20}$$

where $a < 1$ is the internal tube radius. At the elastic-plastic boundary $r = R$, continuity of stresses is assumed:

$$[P] = [S] = [\nu] = 0 \tag{4.21}$$

where [...] denotes a jump of the function. The unknowns will be expanded in series as in Equation (4.15); they are p, S, v and R. In this case, it is more convenient to expand the external load q instead of R:

$$q = q_0 + \delta q_1 + \delta^2 q_2 + O(\delta^2). \tag{4.22}$$

Using Equations (4.12) – (4.14), (4.15), (4.22), (4.20) and (4.21), a solution for the elastic state may be written

$$P_i = \frac{a^2}{1 - a^2} q_i, \; S_i = -\frac{a^2}{r^2}\frac{q_i}{1 - a^2}, \; i = 0, 1, 2, \cdots$$

$$\nu_0 = 0, \; \nu_1 = 1, \; \nu_j = 0, \; j = 2, 3, \cdots \tag{4.23}$$

Initially, plasticity arises at $r = a$ and the load is equal to

$$|q| = |q^*| = |q^*_{(n)}|\left[1 - \delta^2 \frac{2}{3} a^4 + O(\delta^2)\right] \tag{4.24}$$

where $|q^*_{(n)}| = 1 - a^2$ is the corresponding load for an incompressible material. For the elastic-plastic state $|q| > |q^*|$ in the elastic region $R < r < 1$, the following applies

$$P = \operatorname{sign}(q)\left[R^2 - \delta^2 R^6 + O(\delta^2)\right]$$

$$S = -\operatorname{sign}(q)\left[\frac{R^2}{r^2} - \delta^2\frac{2}{3}\frac{R^6}{r^2} + O(\delta^2)\right] \tag{4.25}$$

$$\nu = \delta.$$

The results in the plastic region, $a < r < R$, are given by

$$P = \operatorname{sign}(q)\left[\varsigma_r - \delta^2\left[\frac{r^4}{R^4}\left(\varsigma_r^2 - \frac{1}{3}\varsigma_r + \frac{1}{6}\right)\right.\right.$$

$$\left.\left. - R^4\left(1 - \frac{2}{3}R^2\right) + \frac{1}{3}R^2 - \frac{1}{6}\right] + O(\delta^2)\right]$$

$$S = -\operatorname{sign}(q)\left[1 - \delta^2\frac{2}{3}\frac{r^4}{R^4}\varsigma_r^2 + O(\delta^2)\right] \tag{4.26}$$

$$\nu = \delta\frac{r^2}{R^2} + O(\delta^2)$$

where

$$\varsigma_r = R^2 - 2\ln(R/r).$$

The relation between the load and radius of the plastic region is

$$|q| = 1 - \varsigma_a - \frac{1}{3}\delta^2\left[R^4(3 - R^2) - R^2 + \frac{1}{2}\right.$$

$$\left. - \frac{a^4}{R^4}\left(\varsigma_a^2 - \varsigma_a + \frac{1}{2}\right)\right] + O(\delta^2) \tag{4.27}$$

where $\varsigma_a = R^2 - 2\ln(r/a)$. It may be shown that the expression in the square brackets is always positive so that for a fixed load, the plastic zone radius increases with decreasing Poisson's ratio ν. Full plasticity in the tube ($R = 1$) appears at the load

$$|q^{**}| = |q^{**}_{(n)}|\left[1 - \frac{1}{3}\delta^2\left(a^4(1 + 2\ln a) - \frac{1 - a^4}{4\ln a}\right) + O(\delta^2)\right] \tag{4.28}$$

where $|q^{**}_{(n)}| = -2\ln a$ is the corresponding load for an incompressible material. When q reaches the value q^{**} for a compressible material, it does not imply the loss of load carrying capacity. The tube may still sustain load, $|q| > |q^{**}|$. The state of full plasticity can be presented through the parameter C as

$$P = \operatorname{sign}(q)\left[\varsigma - \delta^2 C^2\left(\frac{1}{6}(r^4 - 1) + r^4\left(\varsigma - \frac{1}{3}\varsigma\right)\right) + O(\delta^2)\right]$$

$$S = -\operatorname{sign}(q)\left[1 - \delta^2\frac{2}{3}C^2 r^4\varsigma^2 + O(\delta^2)\right] \tag{4.29}$$

$$v = \delta C r^2 + O(\delta^2)$$

$$|q| = |q^{**}_{(n)}| \left[1 - \frac{1}{3}\delta^2 C^2 \left\{a^{4a}(1 + 2\ln a) - \frac{1 - a^4}{4\ln a}\right\} + O(\delta^2)\right]$$

where $\varsigma = 1 + 2\ln r$. The parameter C is equal to 1 for $q = q^{**}$ and approaches 0 with increasing load. For $C = 0$, the tube for a compressible material loses the capability to carry a load. The limiting load coincides with the limit q^{**}_n for a tube made from an incompressible material. The radial displacement U_r in the entire cross section of the tube becomes infinity in the limiting state. This can be seen from Equations (4.11), (4.29) and $U_r = r\varepsilon_\phi$.

The results of this section show that compressibility can influence displacement while the limiting loads corresponding to loss of load carrying capacity are the same as for an incompressible material. Moreover, the limiting load does not always depend on the elastic properties of the material.

4.3 Stationary tensile crack in elastic-plastic material

Details of the stress-strain near the tip of a stationary crack is important for formulating crack initiation criterion. Reference can be found in [5, 6]. The work in this section will neglect geometric nonlinearity while the effects of finite deformation and changes in geometry near the crack tip have been treated [7, 8].

Let a crack in an initially stress free material be subjected to monotonically increasing load. The elastic stress field near a crack tip is described by the well-known equations [9]:

$$\sigma_{ij}^{(\alpha)} = \frac{k_\alpha}{\sqrt{2\pi r}} f_{ij}^{(\alpha)}(\phi) + O(1) \tag{4.30}$$

where r and ϕ are the polar coordinates referred to the crack tip in Figure 4.4. k_α are the stress intensity factors and α denotes a particular mode of fracture. For low stress level, small-scale yielding [10] prevails and the crack tip plastic zone is sufficiently small that Equations (4.30) remain valid. Closed form solutions [10, 11] of cracks under longitudinal shear have been obtained for both nonhardening and hardening materials. The same cannot be done in plane extension, in which case only asymptotic or numerical solutuion can be obtained unless the shape of the plastic zone is assumed as *a priori* [12].

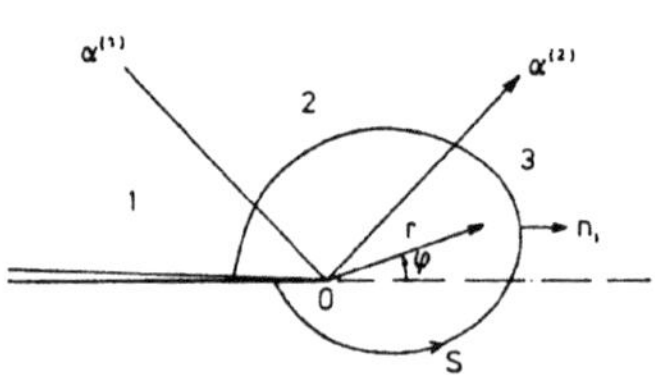

Fig. 4.4: Asymptotic crack tip sectors and contour.

Asymptotic plane strain stress analysis

An asymptotic crack tip stress state as a function of the distances ρ assumes that

$$\rho \ll L, \quad \rho \ll R_p \tag{4.31}$$

where R_p is the maximum size of the plastic zone and L a characteristic length dimension. Solutions of this type have been found [13 – 15] for a linear power hardening material such that $\sigma_{ij}\varepsilon_{ij} \sim r^{-1}$ holds near the crack tip and the angular dependence of the stresses and strains depend on the rate of hardening.

Consider the asymptotic crack tip stress state in an ideal elastic-plastic material which is in a state of plane strain [16]. For any one of the yield conditions in Equation (4.2), the stresses σ_r, σ_ϕ and $\sigma_{r\phi}$ are bounded when $r \to 0$. If, for all ϕ, there exist a finite limit for the radial displacement

$$\lim_{r\to 0} u_2(r, \phi) = U_r(\phi)$$

then for all ϕ, there exist the corresponding limit $U_\phi(\phi)$ for the angular displacement U_ϕ and

$$\lim_{r\to 0} r\varepsilon_r = \lim_{r\to 0} r\varepsilon_\phi = 0$$

$$\lim_{r\to 0} r\varepsilon_{r\phi} = \frac{1}{2}\left(\frac{\mathrm{d}U_r}{\partial \phi} - U_\phi\right). \tag{4.32}$$

Shear strain $\varepsilon_{r\phi}$ is singular, of the order r^{-1}, for all ϕ that satisfy the inequality $\mathrm{d}U_r/\mathrm{d}\phi - U_\phi \neq 0$. This is the highest possible order of strain singularity.

For monotonic loading and the deformation theory of plasticity, the path independent J-integral [10] may be used, Figure 4.4:

$$J = \int_{\mathscr{S}} \left[\int_0^{\varepsilon_{ij}} \sigma_{ij}\, \mathrm{d}\,\varepsilon_{ij}\right] \mathrm{d}x_2 - \sigma_{ij} n_j \frac{\partial u_i}{\partial x_1}\, \mathrm{d}\mathscr{S}. \tag{4.33}$$

From Equations (4.1), (4.32) and (4.33), it follows that

$$J = \lim_{r\to 0} 2 \int_{-\pi}^{\pi} r\varepsilon_{r\phi}(\mathrm{sign}(\varepsilon_{r\phi})\tau_S \cos\phi + \sigma_2 \sin\phi)\, \mathrm{d}\phi. \tag{4.34}$$

In general, the J-integral is nonzero. The continuity of $\varepsilon_{r\phi}$ along ϕ = const. guarantees that at least one angular sector must exist where strain $\varepsilon_{r\phi}$ has a singularity of the order r^{-1}.

Owing to the symmetry of the problem, only the upper half plane $x_2 > 0$ needs to be considered. Near the crack tip, asymptotic expressions for strain with different orders prevail in the sectors $\phi = a^{(i)}$, $i = 1, \ldots, n$ in Figure 4.4. The boundary conditions for a continuous stress field have the form

$$\phi = \pi\colon\ \sigma_\phi = \sigma_{r\phi} = 0$$

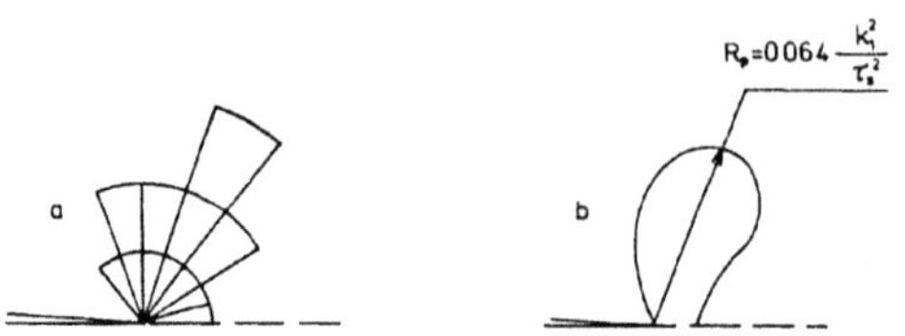

Fig. 4.5: Small-scale yielding: (a) plastic elements and (b) smooth elastic-plastic boundary.

$$\phi = \alpha^{(i)}: \ [\sigma_\phi] = [\sigma_r] = [\sigma_{r\phi}] = [\nu] = 0 \tag{4.35}$$

$$\phi = 0: \ \sigma_{r\phi} = \frac{\partial \sigma_r}{\partial \phi} = 0.$$

Introduce the stress function $\Phi(r, \phi)$ so that Equations (4.12) are satisfied:

$$\begin{aligned} \sigma_2 &= \frac{1}{r^2}\frac{\partial \Phi}{\partial \phi^2} + \frac{1}{2}\frac{\partial \Phi}{\partial r} \\ \sigma_\phi &= \frac{\partial^2 \Phi}{\partial r^2} \\ \sigma_{r\phi} &= -\frac{1}{2}\frac{\partial^2 \Phi}{\partial r \partial \phi} + \frac{1}{r^2}\frac{\partial \Phi}{\partial \phi}. \end{aligned} \tag{4.36}$$

In the sector where $\varepsilon_{r\phi}$ is singular with the order r^{-1} as seen from Equations (4.11) and (4.32), $\nu \to 0$ and $S \to 0$ when $r \to 0$. The zero approximation of the stresses in this sector corresponds to a fan with slip lines radiation from a common point; the sector cannot close the crack flank nor coincide with the line ahead. If the plastic zone completely encloses the crack tip, a zero approximation in r and δ can be determined from Equations (4.2) and (4.35); it consists of three sectors:

$$\begin{aligned} &\alpha^{(1)} = \frac{3\pi}{4}, \ \alpha^{(2)} = \frac{\pi}{4} \\ &\Phi^{(1)} = \frac{r^2}{2}(1 - \cos 2\phi), \ \Phi^{(2)} = \frac{r^2}{2}\left(1 - 2\phi + \frac{3\pi}{2}\right) \\ &\Phi^{(3)} = \frac{r^2}{2}(1 + \cos 2\phi + \pi) \end{aligned} \tag{4.37}$$

where the superscripts identify the sector number in Figure 4.4. The stress field corresponding to Equation (4.37) is the Prandtl solution which was used to describe the crack tip behavior [10, 17]. It is questionable whether the plastic zone should join the crack flank. If the plastic zone does not completely embrace the crack tip, then the stress field may be constructed from the zero approximation term of r and δ in which the first sector enclosed by $\alpha^{(1)}$ would be in the elastic state but not determined [18]. To find the angle $\alpha^{(1)}$, it is necessary to invoke the continuity of ν. According to the work in [18], the material in the first sector in Figure 4.4 is elastic, while the plastic fan prevails in the second sector. For a compressible material, this requires $\nu = \delta = \text{const} \neq 0$ in the first sector and $\nu \to 0$ as $r \to 0$ in the second sector. This condition, however, cannot be satisfied along $\phi = \alpha^{(1)}$; the plastic zone for a compressible material, therefore, completely embraces the crack tip and the Prandtl field in Equation (4.37) corresponds to the zero approximation in r and δ for the asymptotic stress field.

Small scale yielding near crack tip

Static problems of cracks have been solved by using the finite element method (FEM) [19]. Solutions, however, were seldom given in terms of the displacements as a rule and *a priori* assumption is usually made on the nature of the near field strain behavior such that the singularity is embedded into the elements. The elastic-plastic crack problems are solved by making use of the Prandtl–Reuss theory while the load is increased incrementally [20, 21].

Deformation theory has been used to obtain the plastic zone near the crack tip for an ideally elastic-plastic material [22, 23]; the results differed significantly from the corresponding results based on the incremental theory [24]. One of the possible reasons may be attributed to the use of displacement boundary condition instead of the stress condition in Equation (4.30) that are remotely away from the crack. Localized plasticity tends to affect the long distance displacements more than the stresses. It is therefore of interest to solve the small-scale yielding crack problem using the deformation theory with stress boundary conditions given by Equation (4.30). The method of local variations (MLV) [25] will be used. It involves the direct minimization of the finite difference representation of a functional by varying the unknown functions in steps. Since the stresses in an ideal elastic-plastic material are bounded, stress formulation would be the logical choice even for a compressible material.

Apply the variational principle of minimum complementary energy [26]. In plane strain for an elastic-plastic material (4.1), the Mises yield criterion and the stress boundary conditions, it can be shown that in the real state

$$I = \frac{1}{2\mu} \int_{D_e} \left[\frac{2}{3}(1 - 2\nu)(1 + \nu)p^2 + \tau^2 \right] dD + $$

$$+ \int_{D_p} \frac{1}{6}K(3p - \text{sign}\,(p)\sqrt{3(\tau_S^2 - S^2)}\; dD \to \min \tag{4.38}$$

where D_e is the elastic and D_p the plastic portion of the total region D. The minimum of I is to be found for the statically admissible stress fields that satisfy the boundary conditions. The region D is a semicircle with radius R where the origin coincides with the crack tip. The crack flanks are assumed to be traction free while for $r = R$, the stresses σ_r and $\sigma_{r\phi}$ correspond to Equations (4.30). The condition $R/R_p >> 1$ where R_p is the maximum radius of the plastic zone is enforced. The stresses in Equations (4.38) were expressed in accordance with Equation (4.36) for the finite difference approximation of the derivatives of the function Φ.

Figure 4.5 gives the results in the region D which contains 100 elements and $k_1^2/(\tau_s^2 R) = 2.52$. Figure 4.5(a) shows the plastic elements and Figure 4.5(b) the elastic-plastic boundary after smoothing by interpolation. The elastic-plastic boundary is similar to that in [24]. The maximum radius of $R_p = 0.064\, k_1^2/\tau_s^2$ of the plastic zone is situated between $\phi = 60°$ and $80°$ as compared with $R_p = 0.053\, k_1^2/\tau_s^2$ given in [24].

Figure 4.6 shows the stress dependence on ϕ for $r = 0.04\,R$. The stress field presented agrees qualitatively with the Prandtl field while $\sigma_{r\phi}$ agrees quantitatively. This study shows that the deformation and increment theories lead to similar results. The variational principle of complementary energy obtained the near crack tip solution without *a priori* knowledge of the asymptotic stresses. The algorithm of the MLV involving restrictions on the second derivatives is much more complex [27].

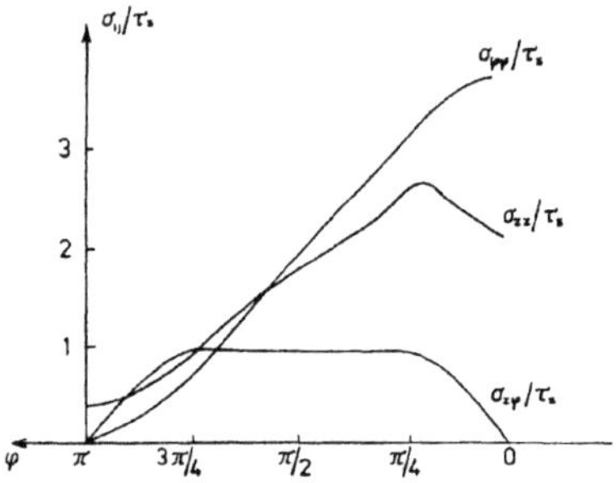

Fig. 4.6: Near field stress dependence on angle ϕ in ideal elastic-plastic material.

Developed plasticity at crack tip

For small-scale yielding, crack initiation can be determined by the critical stress intensity factor k_{1c}. This approach does not apply when plasticity is developed. The path independent J-integral for plane crack problems in linear and nonlinear elasticity may be taken as a parameter to describe the crack tip state of affairs. The criterion $J = J_{1c}$ was suggested for a wide class of materials [28, 29] including the elastic-plastic class. The strain energy density criterion is an example of another approach to the prediction of crack growth under the gross yielding [30]. It was shown in [28] that the J-integral did not depend so much on the local state of the material in the vicinity of the crack tip, i.e., on the hardening behavior

$$\tau = \tau(\gamma), \ \gamma = \sqrt{\frac{2}{3}((\varepsilon_1 - \varepsilon_2)^2 + (\varepsilon_2 - \varepsilon_3)^2 + (\varepsilon_3 - \varepsilon_1)^2)} \tag{4.39}$$

for $\gamma \rightarrow \infty$. Moreover, this behavior cannot be verified experimentally. On the other hand, experimental stress-strain curve (4.39) for small γ should be taken into account with high accuracy when the J-integral is calculated. In [31], J-integral was computed for a compact tension specimen in Figure 4.7 using the deformation theory.

The principle of minimum energy was used [26] to obtain

$$I = \int_D \left(\frac{\varepsilon^2}{6K} + \int \tau(\gamma)\, \mathrm{d}\gamma \right) \mathrm{d}\mathscr{D} - A \rightarrow \min$$

where A is the external work. Minimization was carried out using a modified version of the MLV [31] which decreases the calculation time. The calculations were performed for an aluminum alloy Д 20-1 ($\nu = 0.3$) with the uniaxial tension diagram shown in Figure 4.8. Plastic zone growth up to the fully developed condition were

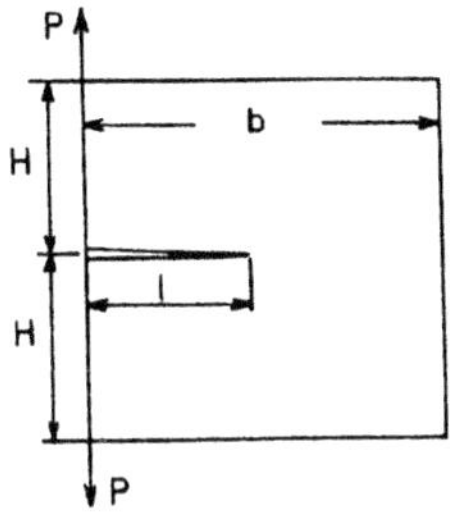

Fig. 4.7: Compact tension specimen.

studied for different geometries. The results confirmed the use of deformation theory. Figure 4.9 shows the curve $J^0 = 2J\mu(1+\nu)/3\tau_s^2 l$ versus load $P = P/\sqrt{3}\tau_s b$ for the specimen with $H/b = 0.6$, $l/b = 0.25$ and $l/b = 0.5$. The number of finite elements used is equal to 256. The dotted lines represent the corresponding results for the elastic material. Note that full plasticity did not develop for both of the load range shown in Figure 4.9.

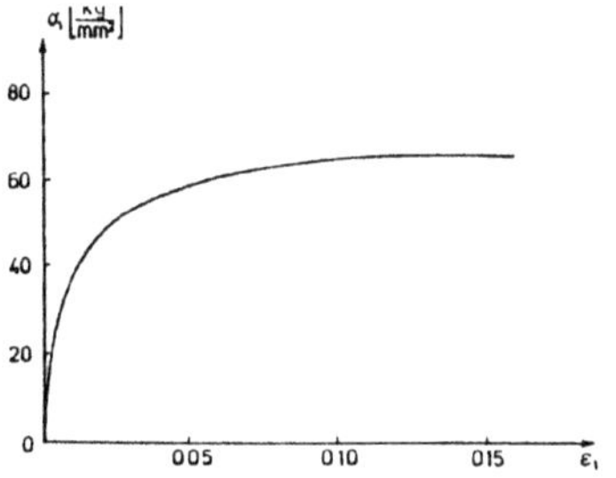

Fig. 4.8: Uniaxial response for aluminum alloy Д 20-1.

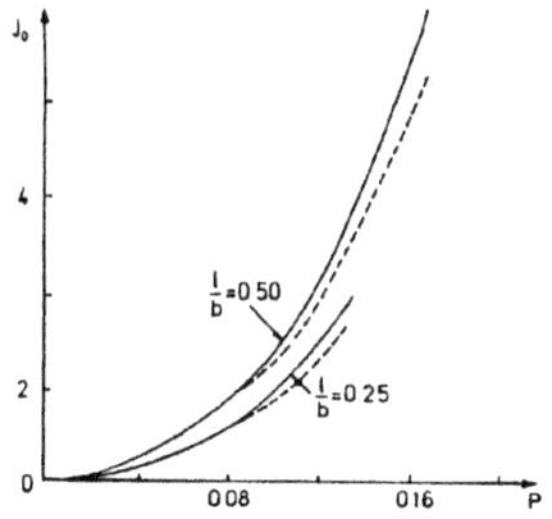

Fig. 4.9: $J^0 = 2J\mu(1+\nu)/(3\tau_s^2 l)$ versus load $p = P/(\sqrt{3}\tau_s b)$ for the aluminum alloy Д 20-1 specimen with $H/b = 0.6$.

4.4 Dynamics of longitudinal shear crack in elastic-plastic medium

Quasistatic and dynamic crack propagation in elastic-plastic materials have been studied in many papers [32 – 39]. Use was made of the incremental and deformation theories of plasticity and the asymptotic stress and strain field. As a result of unloading, there is no input energy into the crack tip for ideally plastic materials [32, 34, 38] and isotropically hardening materials [33, 35, 38]. Some new approaches to the solution of dynamic crack propagation in antiplanar strain for a linearly hardening material will be considered. Discontinuous stress fields are obtained. Necessary conditions for applying the energy approach to fracture are also discussed.

Basic equations

Consider the dynamic crack propagating under longitudinal shear in plane x_1 and x_2:

$$u_i = 0,\ u_3 = W(x_1, x_2, t),\ \gamma_i = \frac{\partial W}{\partial x_i}$$

$$\sigma_{ij} = \sigma_{33} = 0,\ \sigma_{3i} = \tau_i \tag{4.40}$$

$$\frac{\partial \tau_i}{\partial x_i} = \rho \frac{\partial^2 W}{\partial t^2},\ \frac{\partial \gamma_1}{\partial x_2} = \frac{\partial \gamma_2}{\partial x_1}$$

where ρ is the material density, t the time and u_α the displacement vector.

The hardening law is assumed to be isotropic and bilinear with elastic and plastic rigidity μ and μ_p respectively. In what follows only singular stresses and strains are considered. The constitutive equations for loading, unloading and secondary loading may be written as

$$\tau_i = \mu_p \gamma_i,\ \mathrm{d}\tau \geq 0 \tag{4.41a}$$

$$\tau_i = \mu(\gamma_i - \eta\Gamma_i),\ \tau \leq T \tag{4.41b}$$

$$\tau_i = \mu_p \gamma_i + \mu\eta(\Gamma_i^0 - \Gamma_i) + \mu_p \frac{\Gamma - \Gamma_0}{\gamma}(\gamma_i - \Gamma_i^0),\ \tau \geq T,\ \mathrm{d}\tau \geq 0 \tag{4.41c}$$

where T_i and Γ_i are, respectively, the stresses and strains at the beginning of plastic unloading; the superscript 0 denotes the corresponding quantities T_i^0 and Γ_i^0 for secondary loading. Note that $\eta = 1 - (\mu_p/\mu,\ \tau^2 = \tau_i\tau_i,\ \gamma^2 = \gamma_i\gamma_i,\ T^2 = T_iT_i,\ \Gamma^2 = \Gamma_i\Gamma_i$ and $\Gamma_0^2 = \Gamma_i^0\Gamma_i^0$.

Consider a stress free crack propagating along the axis x_1 with a constant velocity c. Introduce the moving coordinates x, y and the polar system r, ϕ in Figure 4.10 such that

$$\frac{\partial}{\partial t} = -c\frac{\partial}{\partial x}. \tag{4.42}$$

The energy g absorbed by the moving crack tip per unit of the crack front may be written as [40]

$$g = \lim_{\varepsilon \to 0} \int_{s_\varepsilon} \left(\left(\int \sigma_{\alpha\beta}\, \mathrm{d}\,\varepsilon_{\alpha\beta} + \frac{1}{2}\rho c^2 \frac{\partial u_\alpha}{\partial x}\frac{\partial u_\alpha}{\partial x} \right) n_1 - \sigma_{\alpha i}\frac{\partial u_\alpha}{\partial x} n_i \right) \mathrm{d}\, s \tag{4.43}$$

where s_ε is an arbitrary contour with a small diameter ε moving with the crack tip and n_i the normal to s_ε in Figure 4.10. Equation (4.43) is valid for an arbitrary constitutive

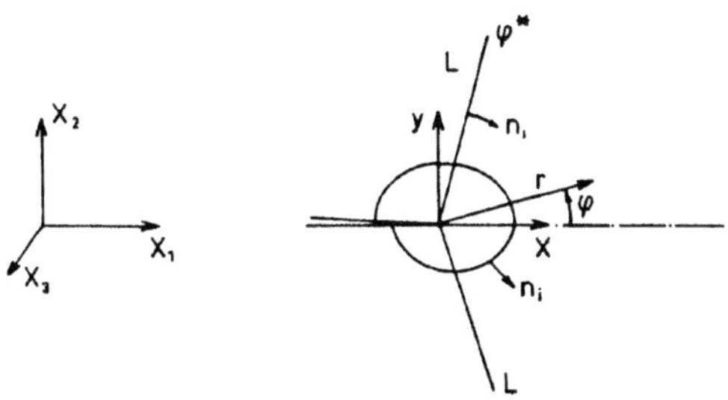

Fig. 4.10: Shock lines L and small contour s_ε near moving crack tip.

law. For anti-plane strain, (4.40) takes the form

$$g = \lim_{\varepsilon \to 0} \int_{s_\varepsilon} \left(\left(\int \tau_i \mathrm{d}\gamma_i + \frac{1}{2} c^2 \gamma_x^2 \right) n_x - (\tau_x n_x + \tau_y n_y) \gamma_x \right) \mathrm{d}s. \tag{4.44}$$

The quantities with subscripts x and y refer to the moving coordinate system. Equation (4.44) is still valid for discontinuous stress fields with finite energy absorption along the shock lines.

Momentum conservation and displacement continuity are satisfied along the shock lines. Equation (4.42) and the condition of small r give

$$[\tau_x n_x + \tau_y n_y - \rho c \gamma_x n_x] = 0 \tag{4.45}$$

$$[\gamma_x n_y - \gamma_y n_x] = 0. \tag{4.46}$$

Asymptotic stress field

Consider the continuous singular stress field near the crack tip moving with a subsonic velocity

$$c < b < a$$

where $a = (\mu/\rho)^{1/2}$ and $b = (\mu_p/\rho)^{1/2}$ are the velocities of the elastic and plastic shear waves, respectively. The velocities c and b will be measured in units of a. Stress conditions and symmetry of the problem suggest that

$$\tau_y = 0, \ \psi = \phi \tag{4.47}$$

$$w = 0, \ \psi = 0 \tag{4.48}$$

With the help of Equations (4.40) to (4.42), the equation of motion may be written as

$$B^2 \frac{\partial^2 w}{\partial x^2} + \frac{\partial^2 w}{\partial y^2} = 0, \ \frac{\partial \tau}{\partial x} \leq 0 \tag{4.49}$$

where

$$B = \sqrt{\left| 1 - \frac{c^2}{b^2} \right|}.$$

For unloading from the plastic state, the equations of motion are of the form

$$A^2\frac{\partial^2 w}{\partial x^2}+\frac{\partial^2 w}{\partial y^2}=\eta\frac{\mathrm{d}\,\Gamma_y(y)}{\mathrm{d}y},\quad \tau\leq T. \tag{4.50}$$

Here, Γ_x, Γ_y, Γ_x^0 and Γ_y^0 are independent of x while $A=\sqrt{1-c^2}$.

The solution of Equations (4.49) and (4.50) is assumed to posses a power type singularity. In general, the state of material in the different sectors near the crack tip may be different. Analysis of solutions of Equations (4.49) and (4.50) shows that, for a continuous field, the order of singularities cannot increase from one sector to another having the greater angle.

Consider the solution for which energy g is finite and nonzero. This condition can be satisfied provided that there exists at least one sector in which γ is of order $r^{-1/2}$. Since the singularity order cannot increase with increasing ϕ, this means that $0\leq\phi\leq\phi^*$ as loading takes place; hence

$$\cos\beta^*\geq\frac{2b^2-c^2-\sqrt{(2b^2-c^2)^2+8c^4}}{4c^2}>-1$$

$$\tan\beta^*=B\tan\phi^*. \tag{4.51}$$

Equation (4.51) implies that the sector of plastic loading can not touch the crack flank and therefore the unloading sector must follow it. The condition of unloading $(\partial\tau)/(\partial x)\,|_{\phi=\phi^*+0}\geq 0$ leads to

$$\cos\beta^*\leq\frac{1+b^2-c^2-\sqrt{(1+b^2-c^2)^2+8c^2(b^2+c^2-1)}}{4c^2}. \tag{4.52}$$

It can be proved that for the continuous asymptotic stress field consisting of the sectors of loading and unloading, the inequalities (4.51) and (4.52) do not permit to satisfy the boundary condition (4.47). Nonzero g does not exist. This conclusion remains unchanged for secondary plastic loading [34, 38].

Consider crack propagation with velocity c that lies between the elastic and plastic case:

$$1>c>b. \tag{4.53}$$

In this case, the material in the sector $0\leq\phi\leq\phi^*$, ahead of the crack is elastic and the solution is nonsingular. Equations (4.40), (4.41a) and (4.42) yield a differential equation of the hyperbolic type:

$$B^2\frac{\partial^2 w}{\partial x^2}-\frac{\partial^2 w}{\partial y^2}=0,\quad \frac{\partial\tau}{\partial x}\leq 0 \tag{4.54}$$

for plastic loading. A singular solution of Equation (4.54) may coexist with the nonsingular elastic solution only along a line of discontinuity. The only possibility to fulfill the condition of plastic loading is to suggest degeneration of the plastic sector onto the line of discontinuity $\phi=\phi^*$ beyond which the state of unloading appears. Since the plastic zone degenerates into a line, only the constitutive equations for unloading are used. Therefore, all results will be valid for any isotropic hardening and for incremental theory of plasticity.

Thus, in the sector $\phi^*\leq\phi<\pi$, unloading from the plastic state takes place and hence Equation (4.50) is valid. Along the shock line and on the crack, the conditions in

Equations (4.45) to (4.47) should be satisfied. The solution in the sector of unloading satisfying the aforementioned conditions is

$$\begin{aligned}
\gamma_x &= \frac{k}{\mu(Ay)^\lambda}(\sin\alpha)^\lambda \cos\Lambda \quad (\tan\alpha = A\tan\phi,\ \Lambda = \lambda(\pi-\alpha)) \\
\Gamma_x &= \frac{kA^2}{\eta\mu(Ay)^\lambda}\frac{\sin(\Lambda^* + \alpha^*)}{(\sin\alpha^*)^{1-\lambda}} \quad (\tan\alpha^* = A\tan\phi^*,\ \Lambda^* = \lambda(\pi-\alpha^*)) \\
\gamma_y &= \frac{kA}{\eta\mu(Ay)^\lambda}\left\{(\sin\alpha)^\lambda \sin\Lambda + \frac{\cos(\Lambda^* + \alpha^*)}{(\sin\alpha^*)^{1-\lambda}}\right\} \\
\Gamma_y &= -\frac{kA}{\eta\nu(Ay)^\lambda}\frac{\cos(\Lambda^* + \alpha^*)}{(\sin\alpha^*)^{1-\lambda}}.
\end{aligned} \tag{4.55}$$

It may be shown that condition of unloading ($\tau \leq T$ for y = *const.*) in the sector $\phi^* \leq \phi < \tau$ can always be satisfied if $0 < \phi_1^* \leq \phi^* \leq \phi^* < \pi$, $0 < \lambda_1 \leq \lambda \leq \lambda_2 < 1/2$, where the values of the parameters ϕ_i^*, λ_i can be found for any c, b and η satisfying the conditions $1 > c > b$, $1 > \eta > 0$. For $\lambda = 1/2$, the total energy absorption along the shock line is infinite, so this case is excluded from consideration. For $\lambda < 1/2$, the above energy absorption is finite, but the energy input to the crack tip is zero ($g = 0$). The detailed investigation of the solution (4.55) and the conditions of its uniqueness may be found in [41, 42].

The case of discontinuous asymptotic stress fields for the stationary moving crack is analogous to that of a continuous one. The energy of deformation released from the elastic part of a body dissipates completely on the shock line in the former case and in the plastic zone in the latter case.

Energy approach to crack propagation in elastic-plastic material

Elastic-plastic problems of crack in stationary motion based on the classical theories of plasticity show that there is no energy input to the crack tip; this means that cracks cannot grow in an elastic-plastic material which is physically unsound. Other models such as that in [35, 38, 43, 44] may have to be considered.

For example, it was proposed in [35] to use the translational hardening model instead of the isotropic one. For very large deformation, the material behaved elastically and energy input prevailed at the crack tip. It may be shown that the more general law of unloading for which plastic or permanent deformation are bounded leads to the same conclusion. It is nevertheless interesting to consider other possibilities for which the classical models give energy input into the crack tip. As mentioned earlier, the stationary crack moving with lines of discontinuity gave no energy input. The other possibility is to investigate the accelerating motion of a crack. There is hope that this latter approach would yield energy input to the crack tip.

4.5 Fracture of viscoelastic solids

Fracture in solids is a time dependent process. Fast crack propagation does not take place right away; it starts only after a certain period of load application. Before that, a slow process of latent fracture occurs even under constant external load. Such a behavior can be accounted for by considering the viscous properties of the material such as linear viscoelasticity.

Crack behavior in viscoelastic solids has been studied in the early papers [40, 45, 46]. Application of the Griffith criterion of fracture to the model without cohesive forces at the crack tip showed that the slow crack growth under constant external load (i.e. crack kinetics) is absent. A crack can either be at rest or propagate dynamically. Many recent papers [47 – 49] have appeared and revealed that crack behavior in viscoelastic solids is still not understood.

Basic equations

The constitutive equations for viscoelastic solids stated in terms of the isotropic $\sigma_{\alpha\alpha}$, $\varepsilon_{\alpha\alpha}$ and deviatoric ($s_{\alpha\beta}$, $e_{\alpha\beta}$) parts of the stress $\sigma_{\alpha\beta}$ and strain $\varepsilon_{\alpha\beta}$ tensors can be written as

$$\sigma_{\alpha\alpha} = 3\frac{1}{K}\{\varepsilon_{\alpha\alpha}\}, \quad s_{\alpha\beta} = 2\mu\{e_{\alpha\beta}\} \tag{4.56}$$

where

$$s_{\alpha\beta} = \sigma_{\alpha\beta} - \frac{1}{3}\delta_{\alpha\beta}\sigma_{\gamma\gamma}, \quad e_{\alpha\beta} = \varepsilon_{\alpha\beta} - \frac{1}{3}\delta_{\alpha\beta}\varepsilon_{\gamma\gamma},$$

$$\frac{1}{K}\{\cdots\} \text{ and } \mu\{\cdots\}$$

stand for the linear integral operators

$$\mu\{f(\tau)\} = \mu_0 f(t) - \int_0^t M(t-\tau)f(\tau)\,d\tau. \tag{4.57}$$

The subscript 0 denotes the instantaneous moduli and compliances while the subscript ∞ denotes the long-term moduli and compliances. For instance

$$\mu_\infty = \mu_0 - \int_0^\infty M(\tau)\,d\tau.$$

The Griffith problem

Consider the Griffith problem of an unbounded plane that contains a cut of length $2l_0$ along the axis x_1 with the crack centered at the origin. Assume that for $t < 0$, there are no stresses nor strains in the solids and at $t = +0$, the stress $\sigma_{22} = q$ is instantaneously applied at infinity and then maintained constant. It is evident that under this condition, the crack length will increase so that, at time t, the crack length will be $2l(t)$ and $\dot{l}(t) = c \geq 0$. The correspondence principle [46, 50] can be applied.

The asymptotic local stress σ_{22} and displacement u_2 at the right tip of the crack are given by

$$\sigma_{22} = ql^{1/2}\mathrm{Re}(2(z - l))^{-1/2}$$
$$u_2 = q\Omega\{l^{1/2}\mathrm{Im}(2(z - l))^{1/2}\}. \tag{4.58}$$

Here, $z = x_1 + ix_2$ and $\Omega\{\cdots\}$ is the linear integral operator in the plane theory of elasticity, Equation (4.57). This operator is expressed in terms of μ and ν for plane strain and plane stress states, respectively, as $\Omega = \mu^{-1}(1 - \nu)$ and $\Omega = \mu^{-1}(1 + \nu)^{-1}$. Using Equations (4.58) and (4.43), the energy density absorbed at the crack tip becomes

$$g = \frac{\pi q^2 l\Omega_0}{2}\left(1 + \frac{1}{\pi}\int_0^t \frac{\Omega(t - \tau)}{\Omega_0}\frac{c(\tau)}{c(t)}\frac{l^{1/2}(\tau)}{l^{1/2}(t)}\Psi(t, \tau)\,\mathrm{d}\tau\right) \tag{4.59}$$

where

$$\Psi(t, \tau) = \lim_{\varepsilon\to 0}\int_{S_\varepsilon} \mathrm{Re}(z - l(t))^{-1/2}\mathrm{Im}(z - l(\tau))^{-1/2}\,\mathrm{d}s.$$

Assume that γ_* is the surface energy associated with crack running at a velocity c; it is characteristic of a given material and can depend on the velocity c, i.e.,

$$g = 2\gamma_*. \tag{4.60}$$

It can be easily shown that the function $\Psi(t, \tau)$ is nonzero and equal to π only when $l(t) = l(\tau)$. Thus, Ψ is equal to π for the crack at rest and equal to zero for the moving crack. As a result, the energy density g increases with time for a stationary crack and would preserve its initial value for a moving crack. A knowledge of Ψ is needed to obtain crack behavior for different loads.

Assume that the load satisfies the following inequality

$$q > \left(\frac{4\gamma_*}{\pi l_0\Omega_0}\right)^{1/2} = q_0. \tag{4.61}$$

Equations (4.59) and (4.60) show that static equilibrium is not possible for $t > 0$ and that the crack propagates dynamically and can never come to rest since the crack becomes increasingly more unstable as it grows.

If $q < q_0$, the crack is initially at rest. But in this case, $l = l_0$ and Equation (4.59) shows that g increases although the stress intensity factor remains constant. It can occur that

$$q < \left(\frac{4\gamma_*}{\pi l_0\Omega_0}\right)^{1/2} = q_\infty. \tag{4.62}$$

Then the crack will be at rest at all times.

If

$$q_\infty < q < q_0 \tag{4.63}$$

then there exist a time $t = t_r$ at which Equation (4.60) becomes valid and the crack starts to move dynamically never coming to rest. The time of fracture delay t_r can be obtained from Equations (4.59) and (4.60):

$$\Omega_0\left(\frac{q_0^2}{q^2} - 1\right) = \int_0^{t_r} \Omega(\tau)\,d\tau. \tag{4.64}$$

If the stress at infinity grows monotonically with time, then the time to fracture can be obtained from

$$\frac{q_0^2}{q(t_r)} = q(t_r) + \frac{1}{\Omega_0}\int_0^{t_r} \Omega(t_r - \tau) q(\tau)\,d\tau. \tag{4.65}$$

A stable crack

Consider the effects of viscosity on the behavior of a crack, which is stable in an elastic solid. The example is the same as before except the load consists of a pair of equal and opposite concentrated forces P applied at the center of the crack. The asymptotic solution takes the form

$$\sigma_{22} = Pl^{-1/2}\mathrm{Re}(2(z - l))^{-1/2}$$
$$u_2 = P\Omega\{l^{-1/2}\mathrm{Im}(2(z - l))^{1/2}\}. \tag{4.66}$$

With the help of Equations (4.66) and (4.59), the energy density becomes

$$g = \frac{\pi P^2 \Omega_0}{2l}\left[1 + \frac{1}{\pi}\int_0^t \frac{\Omega(t - \tau)}{\Omega_0}\frac{c(\tau)}{c(t)}\frac{l(t)}{l(\tau)}\Psi(t, \tau)\,d\tau\right]. \tag{4.67}$$

Assume first that the load is not too great so that

$$P < \left(\frac{4\gamma_* l_0}{\pi\Omega_\infty}\right)^{1/2} = P_\infty. \tag{4.68}$$

Although the right-hand side of Equation (4.67) increases with time, it never reaches the critical value $2\gamma_*$ and the crack will be at rest at all times. If the load is large enough so that

$$P > \left(\frac{4\gamma_* l_0}{\pi\Omega_0}\right)^{1/2} = P_0. \tag{4.69}$$

Then g immediately exceeds $2\gamma_*$ so that

$$l(+0) = \frac{\pi\Omega_0 P^2}{4\gamma_*}. \tag{4.70}$$

The crack then begins to move slowly so that at time $t = +0$ or at some later moment, it has nonzero velocity. According to Equation (4.67), this leads to the contradiction that $l = const$. This presents a peculiar situation where the crack can neither be at rest nor can it move. A possible explanation is that the crack jumps in steps. Initially, it is overdriven and jumps to some other length, probably overshooting its equilibrium state and then comes to rest until it is again overdriven and so on. The amount of overload cannot be found from the analysis [51]. Hence, the crack propagates in a discontinuous step which can be made small to satisfy the equation of motion:

$$l(t) = \frac{\pi P^2}{4\gamma_*}\left[\Omega_0 + \int_0^t \Omega(\tau)\,d\tau\right]. \tag{4.71}$$

This relation is obtained from Equation (4.67) under assumption of no motion. For the range of load

$$P_\infty < P < P_0$$

the crack behavior is similar to that just described, except initially.

In viscoelastic solids, quasistatic crack growth is absent, i.e., a crack can either be at rest or already in fast motion [46, 52]. However, for loads in the range described by Equation (4.63), the crack can be at rest only for some incubation time t_r; hence, in a degenerate fashion, the quasistatic crack growth can exist in a model of fracture of a linear viscoelastic solid without cohesive forces.

4.6. Constitutive instability

According to Figure 4.1, failure behavior of material with microcracks may be described in terms of elastic-plastic constitutive relation [53] where the development of macroscopic defects is identified with the peculiar localized material instability. The approach appears to apply for materials that are initially weak whereby the classical concept of a dominant crack fails to apply. In contrast to [53], an integral instability criterion is proposed with the additional consideration of discreteness so that localization can be proved rather than postulated and its favorable mode can be obtained.

Integral criterion

Consider a system including the body and the self-balanced forces. Let the system be transformed from an equilibrium state to some other neighboring state under the action of applied forces. If there exists a neighboring state, during transition to which the energy is drawn from the system by the applied forces, then the equilibrium state is said to be slightly unstable [54].

Let the elastic-plastic constitutive law be used in the incremental form:

$$\overset{\nabla}{\sigma}_{\alpha\beta} = L_{\alpha\beta\gamma\lambda}\varepsilon_{\gamma\lambda}, \quad L = \begin{cases} L^p & \text{For plastic response} \\ L^e & \text{For unloading} \end{cases} \tag{4.72}$$

where $\overset{\nabla}{\sigma}_{\alpha\beta}$ is the Yaumann increment of the Cauchy stress tensor, $\varepsilon_{\alpha\beta}$ the small virtual strain tensor, $L^e_{\alpha\beta\gamma\lambda}$ the constant stiffness tensor, and $L^p_{\alpha\beta\gamma\lambda}$ the function depending on the load parameter but not of time nor $\varepsilon_{\alpha\beta}$. The instability criterion means that there exists at least one small virtual strain field for which the piecewise-quadratic functional

$$G = \int_D \left(L_{\alpha\beta\gamma\lambda}\varepsilon_{\alpha\beta}\varepsilon_{\gamma\lambda} - \overset{\nabla b}{\sigma}_{\alpha\beta}\varepsilon_{\alpha\beta}\right) \mathrm{d}D \tag{4.73}$$

is nonpositive. Here, $\sigma^b_{\alpha\beta}$ is a fictitious Cauchy stress tensor related to the boundary tractions. If the constraint is produced by the boundary forces, then their action is assumed to suppress instability:

$$\int_D \overset{\nabla b}{\sigma}_{\alpha\beta}\varepsilon_{\alpha\beta}\, \mathrm{d}D \leq 0. \tag{4.74}$$

The system can become unstable only due to the instability of the material itself, i.e., due to nonpositiveness of the constitutive part of the integrand $L_{\alpha\beta\gamma\lambda}\, \varepsilon_{\alpha\beta}\, \varepsilon_{\gamma\lambda}$. It can take place only for the plastic response because the quadratic form $L^e_{\alpha\beta\gamma\lambda}\, \varepsilon_{\alpha\beta}\, \varepsilon_{\gamma\lambda}$ is definitely positive for unloading.

In addition to the instability criterion, it is necessary to introduce a selection principle for identifying the mode that is more likely to cause instability. Let the load process be determined by a monotonically increasing parameter. The instability mode is assumed to be primary if it corresponds to the least value of this parameter. The constitutive law is [53]

$$L^p_{\alpha\beta\gamma\lambda}\varepsilon_{\gamma\lambda} = 2\mu e_{\alpha\beta} + \frac{\varepsilon}{3K}\delta_{\alpha\beta} - \frac{3\mu K S_{\alpha\beta} + \beta\delta_{\alpha\beta}}{3K(\mu + h) + m\beta}\left(\mu\xi + m\frac{\varepsilon}{3K}\right)$$

$$L^e_{\alpha\beta\gamma\lambda}\varepsilon_{\gamma\lambda} = 2\mu e_{\alpha\beta} + \frac{\varepsilon}{3K}\delta_{\alpha\beta} \tag{4.75}$$

where $e_{\alpha\beta}$ is the deviatoric strain tensor, $S_{\alpha\beta}$ the normalized deviatoric stress tensor ($S_{\alpha\beta}S_{\alpha\beta} = 2$), h is the plastic hardening modulus, m is the internal friction coefficient and β is the dilatancy factor. In Equation (4.75), $\varepsilon = \varepsilon_{\alpha\alpha}$ and $\xi = e_{\alpha\beta}S_{\alpha\beta}$. The condition of plastic response is

$$\xi^p = \frac{3\mu K\xi + m\varepsilon}{3K(\mu + h) + m\beta} \geq 0 \tag{4.76}$$

Figure 4.11 shows the plastic shearing curve which is assumed to be convex, i.e., the plastic modulus decreases monotonically with deformation. The body undergoes uniform plastic loading up to the onset of instability, when a nonuniform strain distribution arises. Hence, for the state under consideration, the plastic modulus h distribution is uniform. The value of $(-h)$ is adopted [55] as a parameter involved in the selection principle.

Note that the constitutive law is nonlinear due to discontinuity in the modulus. This nonlinearity together with constraint cause localization. A simple case of constitutive instability corresponds to a nonconstrained body. In this case, Equation (4.73) reduces to degeneration of the quadratic form $L^p_{\alpha\beta\gamma\lambda}(h)\varepsilon_{\alpha\beta}\varepsilon_{\gamma\lambda}$. The critical values are

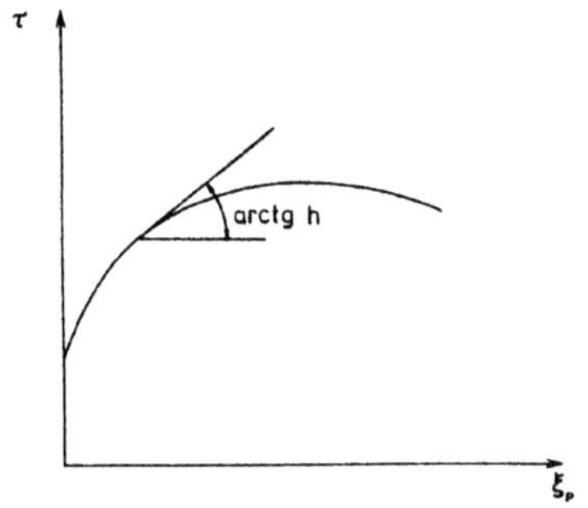

Fig. 4.11: Plastic shearing curve for material with internal friction and volume change.

$$h_0 = \frac{(m - \beta)^2}{12K} + O(\alpha^4)$$

$$\varepsilon^0_{\alpha\beta} = \left[\left(\frac{\alpha}{3} + O(\alpha^3)\right)\delta_{\alpha\beta} + \frac{1}{2}S_{\alpha\beta}\right]\xi^0, \; \xi^0 > 0 \tag{4.77}$$

where $m = \alpha m_0$, $\beta = \alpha\beta_0$, $1/2(m_0 + \beta_1) = 1$ and $0 < \alpha < 1$.

Equation (4.77) shows that the instability of a nonconstrained body is accompanied by a volume increase. If the volume increase is impossible, then this type of instability will not arise. Constitutive instability occurs only for $h \leq h_0$ with $h = h_0$ being a limiting case.

Localization of critical strain in a constrained body

Consider two different types of constraint: kinematic and stiffness. Kinematic constraint implies that some restrictions are imposed on the boundary displacement:

$$\int_{\mathscr{S}} u_\alpha n_\alpha \, d\mathscr{S} = 0 \Leftrightarrow \int_D \varepsilon \, dD = 0.$$

Stiffness constraint implies that the inequality in Equation (4.74) is imposed. The application of elastic clamp is an example of a stiffness constraint; it becomes kinematic when the clamp stiffness tends to infinity. In the absence of constraint, instability would corrrespond to the virtual strain field $\varepsilon_{\alpha\beta} = \xi\varepsilon^0_{\alpha\beta}$, where ξ is a constant or a linear function of the coordinates. With constraint, such strains are not allowed at kinematic constraint and will not lead to instability for stiffness constraint.

There are two different ways for instability to occur under constraint. For simplicity, consider kinematic constraint. The first one corresponds to uniform distribution of virtual strain compatible with constraint but differ greatly from $\xi\varepsilon^0_{\alpha\beta}$. The second one involves highly nonuniform distribution with localized strain close to $\xi\varepsilon^0_{\alpha\beta}$ in a zone of small volume with relatively small strains outside this zone. Qualitative estimates conclude that localizational instability can occur only if the zone tends to a thin layer. More complete localization involving the smallness of two or three dimensions of the zone cannot yield instability.

In the above mentioned estimates, continuity of the virtual strain components tangent to the zone of localization was not taken into account. Such consideration is adequate for one dimension where no tangent components occur and the exact solution of the problem for both types of constraint [52] gives a δ-like virtual strain distribution, i.e., complete localization. The latter can also occur for arbitrarily small constraint stiffness in two dimensions, but not for three dimensions.

Localization in three dimensions can be illustrated by two model problems [55 – 57]. In the first, the critical parameters are determined under the condition that instability is characterized by complete localization in extremely thin layers. The analysis of the functional $G(h, \varepsilon_{\alpha\beta})$ is then reduced to the analysis of the integrand $L^{p}_{\alpha\beta\gamma\lambda}(h)\varepsilon_{\alpha\beta}\varepsilon_{\gamma\lambda}$ under the condition that there exists a zero plane for the tensor $\varepsilon_{\alpha\beta}$, because its components tangent to the localization layer surface must be zero. The critical values obtained are $h_* < h_0$ and $\overset{*}{\varepsilon}_{\alpha\beta} = 1/2(n_\alpha g_\beta + n_\beta g_\alpha)$. For the deviator of pure shear $S_{\alpha\beta} = \delta_{1\alpha}\delta_{1\beta} - \delta_{2\alpha}\delta_{2\beta}$ we have:

$$
\begin{aligned}
n_\alpha &= \cos\left(\frac{\pi}{4} - \psi\right)\delta_{1\alpha} + \sin\left(\frac{\pi}{4} - \psi\right)\delta_{2\alpha} \\
g_\alpha &= \cos\left(-\frac{\pi}{4} + \psi\right)\delta_{1\alpha} + \sin\left(-\frac{\pi}{4} + \pi\right)\delta_{2\alpha} \qquad (4.78) \\
\psi &\cong \frac{1}{2}\arcsin\left(\frac{\alpha}{\mu K + 1}\right).
\end{aligned}
$$

The two likely planes of localization planes are orthogonal to n_α and g_α, respectively.

While the character of deformation for complete localizaiton is revealed in the first problem, the favorable condition for localizational instability is not determined. The second model will study the situation in search for localization as a function of constraint stiffness. The body is assumed to be a plane layer with unit thickness and elastic clamp stiffness c for varying layer thickness as shown in Figure 4.12. The boundary planes are parallel to one of the two favorable planes of localization. The zone of localization is assumed to be a plane layer of thickness a, where $0 \leq a \leq 1$. The critical value $a = a_*(c)$ corresponds to a threshold constraint stiffness c_* which is magnitude of the elastic modulus:

$$
\begin{aligned}
&0 \leq c \leq c_* \Rightarrow a_* = 1 \\
&c = c_* \Rightarrow a_* \text{ is arbitrary between 0 and 1} \qquad (4.79) \\
&c > c_* \Rightarrow a_* = 0.
\end{aligned}
$$

Simultaneous variations of the orientation and thickness of the localized zone can also be examined. Localization occurs in the form of a δ-like distribution of dyadic strain given in Equation (4.78). Localizational instability is not always the favorable mode of constitutive instability; it occurs only when the constraint is sufficiently stiff, Equation (4.79).

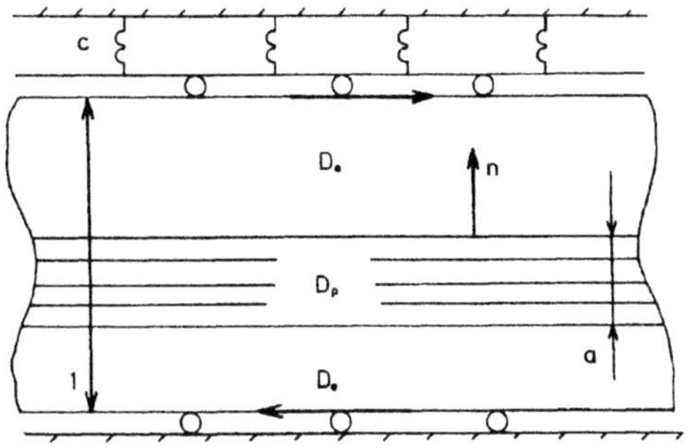

Fig. 4.12: Localization in constrained body.

Fault systems

Since all localizational instability modes correspond to the same critical value h_*, discreteness of the material must be considered to eliminate this degeneration. It is postulated that localization layer thickness cannot be less than a certain dimension d_0 that is inherent in the material. The instability modes are considered to be almost completely localizational. The selection principle for them takes the modified form based on perturbation procedure with small parameter $\delta = d_0/L$ with L being a characteristic length dimension [55].

$$h = h_0 - \delta h_1 + O(\delta)$$

$$h_1 = \frac{3\mu K + (\alpha^2/\mu K + 1)}{3K(\mu + h_*) + m\beta} \frac{\int_{De} L^e_{\alpha\beta\gamma\lambda}\varepsilon_{\alpha\beta}\varepsilon_{\gamma\lambda}\,\mathrm{d}D}{\int_{\omega_p} \xi^2\,\mathrm{d}\omega} \tag{4.80}$$

$$-h_1 = \min \Leftrightarrow \Lambda \equiv \frac{\int_{De} L^e_{\alpha\beta\gamma\lambda}\varepsilon_{\alpha\beta}\varepsilon_{\gamma\lambda}\,\mathrm{d}D}{\int_{\omega_p} \xi^2\,\mathrm{d}\omega} = \min.$$

Here, ω_p is the union of middle-planes of the localization layers and ξ is the localized strain amplitude distribution. The functional Λ governs the geometric similarity of the localization instability patterns; when it is invariant under translation, it governs also the equal spacing of localization layers. Equation (4.80) will be used to estimate a qualitative feature of the fault systems which corresponds to the shear of a square box and a strip sandwiched between rigid walls. In both cases, the finite-element method is used to determine the kinematics in the unloading region and the corresponding piecewise-linear distribution of ξ along the localization lines. Figure 4.13 shows the parallel localization lines equally spaced at αL; they are obtained in [55 – 57] in the case of a box. For the strip (Figure 4.14), localization lines of two possible directions (almost parallel and almost perpendicular to the strip axis) are not equivalent: the former are $2/\alpha$ times longer and, hence, are favorable, if the lines of a single type are considered. But in general the combination of both types may appear to be absolutely

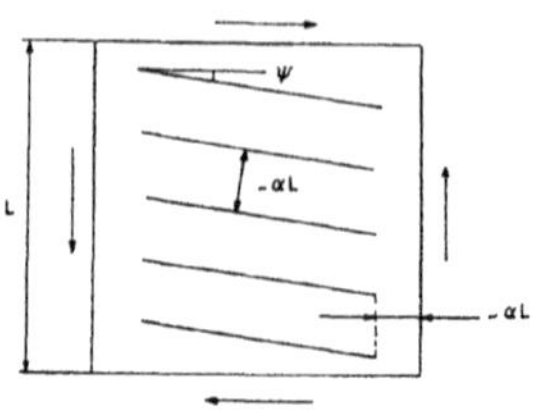

Fig. 4.13: Localization lines for a box.

favorable, and this possibility should also be investigated.

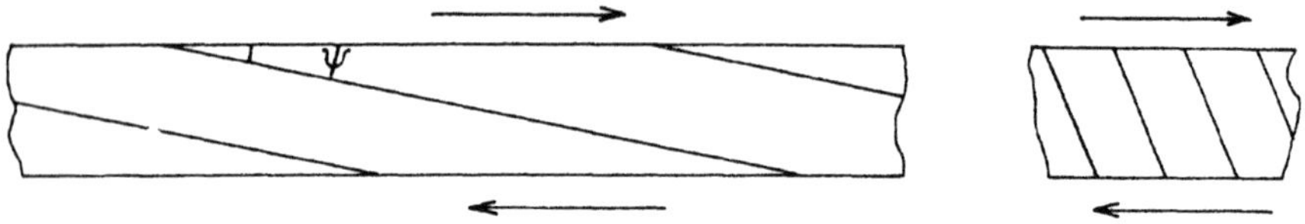

Fig. 4.14: Localization lines for a strip.

Under the conditions of a strip, sandwiched between rigid walls, the above mentioned possibility has not been realized. The 'echelon' fault system formed only by the long lines overlapping approximately by a half of their length (Figure 4.14) has proved to be favorable. These results agree qualitatively with the experiments in [55, 56].

References

[1] RABOTNOV, Yu.N., *Mechanics of Deformable Solids,* Moscow, Nauka (in Russian) 1979.

[2] SOKOLOVSKY, V.V., *Theory of Plasticity,* Moscow, Vischaya shkola (in Russian) 1969.

[3] MUKHAMEDIEV, Sh.A. and NIKITIN, L.V., *Izvestiya AN SSSR. Mechanika tverdogo tela,* No. 2 (in Russian), 1981, pp. 39–47.

[4] IVLEV, D.L. and ERSHOV, L.V., *Perturbation Method in the Theory of Elastic-Plastic Body,* Moscow, Nauka (in Russian) 1978.

[5] VITVITSKY, P.M., PANASJUK, V.V. and IAREMA, S.I.A., *Problemi prochnosti,* No. 2, 1973, pp. 3–18.

[6] RICE, J.R., *The Mechanics of Fracture, AMD,* Vol. 19, 1977, pp. 23–53.

[7] RICE, J.R. and JONSON, M.A., The role of large crack tip geometry changes in plane strain fracture, in *Inelastic Behavior of Solids,* Kanninen *et al.,* eds., McGraw-Hill, 1970, pp. 641–672.

[8] McMEEKING, R.M., *J. Mech. Phys. Solids,* Vol. 25, No. 5, 1977, pp. 357–381.

[9] PARIS, P.C. and SIH, G.C., Analysis of stress state near crack (transl. in Russian). *Prikladnie voprosy vjaskosty razruchenia,* Moscow, Mir, 1968.

[10] RICE, J.R., Mathematical Analysis in the Mechanics of Fracture, in *Fracture: An Advanced Treatise,* H. Liebowitz Ed., Academic Press, Vol. 2, 1967, pp. 191–311.

[11] HULT, J.A. and McCLINTOCK, F.A., Elastic-plastic stress and strain distribution around sharp

notches under repeated shear, *Int. Congr. for Appl. Mech.*, Brussels, Vol. 8, 1957, pp. 51–58.
[12] DUGDALE, D.S., *J. Mech. Phys. Solids,* Vol. 8, No. 2, 1960, pp. 100–104.
[13] CHEREPANOV, G.P., *Prikladnaya matematika i mechanika,* Vol. 31, No. 3, 1967, pp. 476–488 (in Russian).
[14] RICE, J.R. and ROSENGREN, J., *J. Mech. Phys. Solids,* Vol. 16, No. 1, 1968, pp. 1–12.
[15] HUTCHINSON, J.W., *J. Mech. Phys. Solids,* Vol. 16, No. 1, 1968, pp. 13–31.
[16] MUKHAMEDIEV, Sh.A., Stresses and strains under plane strain of nonhardening elastic-plastic material. Application to Fracture (in Russian), VINITI, No. 1188–74, 1974, DEP.
[17] CHEREPANOV, G.P., *Mechanics of Brittle Fracture* (in Russian), Moscow, Nauka, 1974.
[18] IBRAGIMOV, V.A. and TARASJUK, N.E., *Izvestiya AN SSSR. Mechanika tverdogo tela,* No. 5, 1976, pp. 184–185.
[19] ZIENKIEWICZ, O.C. and CHEUNG, Y.K., *The Finite Element Method in Structural and Continuum Mechanics,* McGraw-Hill, 1967.
[20] YAMADA, Y., YOSHIMURA, N. and SAKURAI, T., *Int. J. Mech. Sci.,* Vol. 10, No. 5, 1963, pp. 121–135.
[21] MARCAL, P.V. and KING, I.P., *Int. J. Mech. Sci.,* Vol. 9, No. 3, 1967.
[22] KUDRAJAVTZEV, V.A., PARTON, V.Z., PESKOV, Ju.A. and CHEREPANOV, G.P., *Izvestiya AN SSSR. Mechanika tverdogo tela,* No. 5, 1970, pp. 65–74 (in Russian).
[23] PARTON, V.Z. and MOROZOV, E.M., *Mechanics of Elastic-Plastic Fracture* (in Russian), Moscow, Nauka, 1974.
[24] LEVY, N., MARCAL, P.V., OCTERGREN, W.J. and RICE, J.R., *Int. J. Fract. Mech.,* Vol. 7, No. 2, 1971, pp. 143–156.
[25] CHERNOUSKO, F.L. and BANICHUK, N.V., *Variational Problems in Mechanics* (in Russian), Moscow, Nauka, 1973.
[26] KACHANOV, L.M., *Foundations of Theory of Plasticity* (in Russian), Moscow, Nauka, 1973.
[27] MUKHAMEDIEV, Sh.A. and NIKITIN, L.V., *Prikladnaya matematika i programmirovanie, Kishinev, Shtiintsa,* No. 12, 1975, pp. 92–108.
[28] BROBERG, K.B., *J. Mech. Phys. Solids,* Vol. 19, No. 6, 1971, pp. 407–418.
[29] LANDES, J.D. and BEGLEY, J.A., *J*-integral as a failure criterion, *5th National Symposium on Fracture Mechanics,* University of Illinois, 1971.
[30] SIH, G.C., and MADENCI, E., *Eng. Fracture Mech.,* Vol. 18, No. 3, 1983, pp. 667–677.
[31] MUKHAMEDIEV, Sh.A., NIKITIN, L.V. and YUNGA, S.L., *Isvestiya AN SSSR. Mechanika tverdogo tela.,* No. 1, 1976, pp. 76–83 (in Russian).
[32] ACHENBACH, J.D. and DUNAEVSKY, V., *J. Mech. Phys. Solids,* Vol. 29, No. 4, 1981, pp. 283–303.
[33] AMAZIGO, J.C. and HUTCHINSON, J.W., *J. Mech. Phys. Solids,* Vol. 25, No. 2, 1977, pp. 81–93.
[34] CHITALEY, A.D. and McCLINTOCK, F.A., *J. Mech. Phys. Solids,* Vol. 19, No. 3, 1971, pp. 147–163.
[35] IBRAGIMOV, V.A., *Prikladnaya matematika i mechanika,* Vol. 40, No. 2, 1976, pp. 337–345 (in Russian).
[36] RICE, J.R., Elastic-plastic crack growth, MRL E-124, Brown University, 1980.
[37] SLEPJAN, L.I., *Izvestiya AN SSSR Mechanika tverdogo tela,* No. 1, 1974, pp. 57–67 (in Russian).
[38] SLEPJAN, L.I., *Mechanics of Cracks,* (in Russian), Leningrad, Sudostroenie, 1981.
[39] YU-CHEN, G. and KEH-CHIN, H., Elastic-plastic fields in steady crack growth in a strain-hardening material, *Advances in Fracture Research,* D. Francois Ed., Pergamon Press, Vol. 2,

1981, pp. 669 – 682.

[40] KOSTROV, B.V., NIKITIN, L.V. and FLITMAN, L.M., *Izvestiya AN SSSR. Mechanika tverdogo tela,* No. 3, 1969, pp. 112 – 125 (in Russian).

[41] MUKHAMEDIEV, Sh.A. and NIKITIN, L.V., *Izvestiya AN SSSR. Mechanika tverdogo tela,* No. 6 (in Russian), 1988.

[42] KONDAUROV, V.I., MUKHAMEDIEV, Sh.A., NIKITIN, L.V. and RYZHAK, E.I., *Rock Fracture Mechanics,* Moscow, Nauka (in Russian), 1987.

[43] SIH, G.C., *Theoretical Applied Fracture Mech.,* Vol. 4, No. 3, 1985, pp. 157 – 173.

[44] SIH, G.C. and MACDONALD, B., *Eng. Fracture Mech.,* Vol. 6, 1974, pp. 361 – 386.

[45] WILLIAMS, M.L., The fracture of visco-elastic material, *Fracture of Solids,* Interscience Publishers, N.Y., 1963, pp. 157 – 188.

[46] KOSTROV, B.V., NIKITIN, L.V. and FLITMAN, L.M., *Izvestiya AN SSSR, Fizika Zemli,* No. 7, 1970, pp. 20 – 35 (in Russian).

[47] CHRISTENSEN, R.M., *Int. J. Fracture,* No. 1, 1979, pp. 3 – 21.

[48] McCARTNEY, L.N., *Int. Journal of Fracture,* No. 13, 1977, pp. 641 – 654.

[49] POPELAR, C.H. and ATKINSON, S., *J. Mech. Phys. Solids,* Vol. 28, No. 2, 1980, pp. 79 – 93.

[50] GRAHAM, G.A., *Quart. Appl. Math.,* No. 25, 1968, pp. 167 – 185.

[51] MOLCHANOV, A.E. and NIKITIN, L.V., *Izvestiya AN SSSR. Mechanika tverdogo tela,* No. 2, 1972, pp. 181 – 194.

[52] KOSTROV, B.V. and NIKITIN, L.V., *Archiwum Mechaniki Stosowaney,* Vol. 6, No. 22, 1970, pp. 749 – 775.

[53] RUDNICKI, J.W. and RICE, J.R., *J. Mech. Phys. Solids,* Vol. 23, No. 6, 1975, pp. 371 – 394.

[54] DRUCKER, D.C., *J. Appl. Mech.,* Vol. 26, No. 1, 1959, pp. 101 – 106.

[55] RYZHAK, Ye.I., *Izvestiya AN SSSR. Mechanika tverdogo tela,* No. 5, 1983, pp. 127 – 136.

[56] NIKITIN, L.V. and RYZHAK, Ye.I., *Doklady AN SSSR,* Vol. 230, No. 5, 1976, pp. 1203 – 1206.

[57] NIKITIN, L.V. and RYZHAK, Ye.I., *Izvestiya AN SSSR, Fizika Zemli,* No. 5, 1977, pp. 22 – 37.

[58] REVUZHENKO, A.F., STAZHEVSKY, S.B. and SHEMYAKIN, E.I., *Fiziko-technicheskie problemy,* No. 3, 1974, pp. 130 – 133.

[59] BYERLEE, J., MYACHKIN, V., SUMMERS, R. and VOEVODA, O., *Tectonophys.,* Vol. 44, NO. 1 – 4, 1978, pp. 161 – 171.

G.C. Sih and D.Y. Tzou

5

Plastic deformation and crack growth behavior

5.1 Introduction

Even though the theory of plasticity has been used extensively to describe the permanent deformation of solids, some of the hidden restrictions imposed by the underlying assumptions are subtle and can be unwittingly overlooked in application. Generally speaking, the theory can be divided into two ranges. In situations where the plastic strains dominate the global behavior of the structure, the theory has been successful, as in the case of limit analysis. The same cannot be said for the application of plasticity to regions near holes or defects where the local strain rates can differ appreciably from those at distances away. Ordinary plasticity does not account for any change in material properties with load or strain rate history because the same constitutive relation is generally applied to all elements in a solid body. No provisions are made for the elevation of strain rates in elements next to the defects or cracks. Unless certain modifications are made, plasticity will not give a realistic prediction of crack growth behavior, a problem that has attracted the attention of many past investigations in fracture mechanics.

It is not the objective of this contribution to survey the role of plasticity in fracture mechanics, but rather to investigate the ways with which plastic deformation affects crack growth, an area that has been discussed but not explored in detail. Aside from the limitations of plasticity theory mentioned earlier, too much emphasis cannot be placed on the selection of an appropriate criterion that can consistently describe the nonlinear behavior of crack growth. The segment of each growth step is not constant but depends on the rate at which energy is dissipated during the process of new surface creation [1]. As the material properties ahead of the crack change with load, the condition that determines crack growth alters accordingly. A realistic account of this behavior was first made in [2]; the nonlinear crack growth behavior was predicted from the uniaxial data by application of the volume energy density criterion which is more forgiving than those criteria based on stress or strain quantities. When modelling the material with an elastic-plastic behavior, the order of the crack tip stress or strain singularity tends to depend on the constitutive equation chosen. In other words, the stress and strain field become nonhomogeneous and the idea of a plastic stress or strain intensity factor [3, 4] will no longer be adequate for describing the state of affairs near the crack front. On the other hand, the $1/r$ character of the strain energy density function, dW/dV, will remain unchanged independent of the constitutive relation of the material, where r is the radial distance measured from the crack tip. This unique feature of dW/dV is unmatched by the other criteria. The coefficient of $1/r$ known as the strain energy density factor S [5] can thus provide a unique representation of the elastic-plastic crack tip energy intensity. Predictions on crack growth based on the S-factor can be found in [6 – 8]. An important result is that the rate change of S with crack length, say a, turns out to be linear, i.e., dS/da = const. Changes in crack growth due to differences in specimen sizes and loading rates can thus be predicted. Refer to the results in [9] for cracked specimens in two dimensions while the three dimensional cases can be found in [10]. Similar findings are obtained for fatigue crack growth [11 – 14].

5.2 Classical theory of plasticity

This section provides a very brief outline of the equations in plasticity that will be used in the subsequent example on crack growth, most of which can be found in two of the well-known texts on plasticity [15, 16]. It is customary to assume that a hydrostatic stress will not significantly affect the yield strength of metals* and hence, it is expedient to define the deviatoric stress

$$\mathcal{S}_{ij} = \sigma_{ij} - \frac{1}{3}\sigma_{kk}\delta_{ij}, \quad i, j = x, y, z. \tag{5.1}$$

Dilatational effects are assumed to have no influence on the ways with which solid elements distort. Yielding of an element in a multiaxial state of stress, therefore, is assumed to depend only on the effective stress

$$\sigma_{eff} = \frac{1}{\sqrt{2}}[(\sigma_{xx} - \sigma_{yy})^2 + (\sigma_{yy} - \sigma_{zz})^2 + (\sigma_{xx} - \sigma_{zz})^2 + 6(\sigma_{xy}^2 + \sigma_{yz}^2 + \sigma_{xz}^2)]^{1/2} \tag{5.2}$$

which corresponds to the distortional component of the elastic strain energy density function. Loading into the plastic range is said to occur when

$$\sigma'_{eff} \geq \sigma_{yd} \quad \text{and} \quad d\sigma'_{eff} \geq 0 \tag{5.3}$$

where σ_{yd} stands for the yield stress in the uniaxial test. The prime refers to the center of the current yield surface in the stress space.

Isotropic and kinetic hardening

A material is said to be plastically hardened when the rate increase in strain exceeds that in stress. This behavior is modelled in plasticity by referring to the topological variation of the yield surface in terms of kinematic and isotropic hardening. Assumed in both situations is that the three dimensional yield stress surface moves with the

* The early experiment of Bridgman [17] showed that the yield strength is not greatly affected for hydrostatic pressure up to 1.4×10^5 psi.

current stress deviator $\mathrm{d}\mathscr{S}_{ij}$. Moreover, the incremental change $\mathrm{d}C_{ij}$ of the coordinates of the center of the yield surface is taken to be proportional to the projection of the current total stress deviator onto the local normal of yield surface. The corresponding mathematical expressions are

$$\mathrm{d}C_{ij} = \begin{cases} \dfrac{3}{2(\sigma'_{eff})^2}\lambda\, \mathscr{S}'_{kl}\,\mathrm{d}\mathscr{S}'_{kl}\,\mathscr{S}'_{ij}; & \text{for}\,\sigma'_{eff} \geq \sigma_{yd} \text{ and } \mathrm{d}\sigma'_{eff} \geq 0 \\ 0; & \text{otherwise.} \end{cases} \tag{5.4}$$

The parameter λ controls the hardening characteristics; a value zero represents isotropic hardening such that shape of the yield surface is self-similar as illustrated in Figure 5.1(a) and a value of one denotes kinematic hardening where the yield surface undergoes a rigid body translation as shown in Figure 5.1(b). Any value of λ between zero and one corresponds to general hardening and gives rise to yield characteristics that combine isotropic and kinematic hardening. The corresponding material true stress and true strain curves are given in Figures 5.2(a) and 5.2(b). Note that isotropic hardening assumes a higher yield stress upon reloading while kinematic hardening maintains the same yield stress but relocates the yield strain as a result of a rigid body translation of the stress and strain curve. Such characteristics, once assumed, are intrinsic of the material behavior at every location and invariant with reference to the load history. Hence, the end results in the theory of plasticity are directly influenced by the selection of flow rule, a choice that is not obvious.

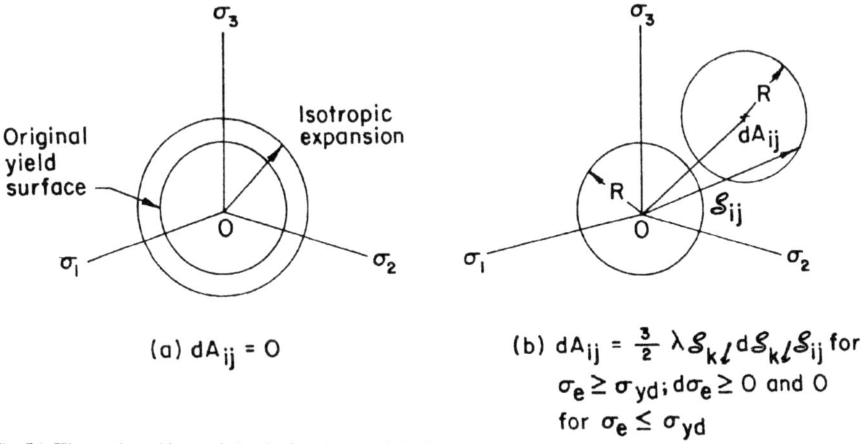

Fig. 5.1: Kinematic and isotropic hardening characteristics in the π-plane.

Another major consideration in plasticity is the ways with which elastic strain ε^e_{ij} and plastic ε^p_{ij} would interact. It is assumed that the total strain can be decomposed linearly into two parts

$$\varepsilon_{ij} = \varepsilon^e_{ij} + \varepsilon^p_{ij}. \tag{5.5}$$

The possibility of any nonlinear interaction is not considered. It follows from Equation (5.5) that the incremental change becomes

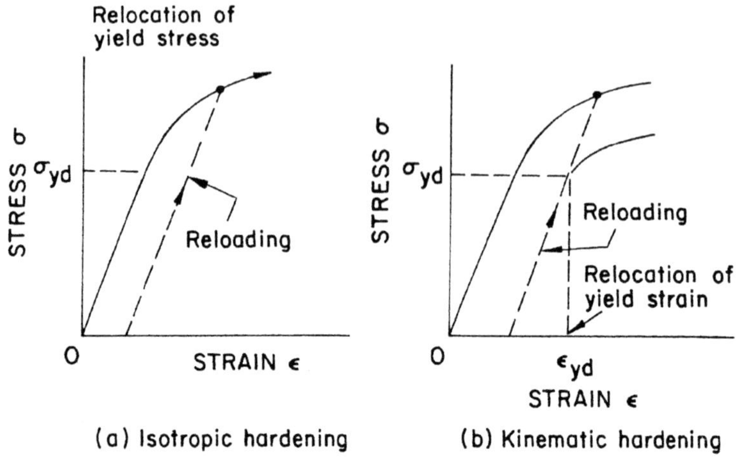

Fig. 5.2: Material true stress and true strain curves.

$$d\varepsilon_{ij} = d\varepsilon^e_{ij} + d\varepsilon^p_{ij}. \tag{5.6}$$

Path of deformation

The deformation theory of plasticity makes use of the secant modulus to distinguish the current and reference states of deformation as shown in Figure 5.3. The path from 0 to A does not affect the final state (σ_s, ε_s): the resulting deformation is said to be path independent. Path dependency comes into play in the incremental theory of plasticity where the stress and strain states between B and A in Figure 5.3 will depend on the tangent modulus. The governing equation is that of Prandtl–Reuss [18]:

$$d\varepsilon^p_{ij} = d\lambda(\sigma_{eff})\mathscr{S}'_{ij}, \quad \text{for} \quad \sigma'_{eff} \geq \sigma_{yd} \quad \text{and} \quad d\sigma'_{eff} \geq 0. \tag{5.7}$$

It assumes that the principal axes of the incremental plastic strain coincides with those of the total stress deviator and is proportional to the total stress deviator. The incremental proportionality factor $d\lambda$ in Equation (5.7) depends on the local tangent modulus at a given strain in the plastic range. Plastic flow is also assumed to be incompressible, i.e., $\nu = 0.5$, and hence

$$d\lambda = 3f(\sigma_{eff})d\sigma_{eff} \tag{5.8}$$

with $f(\sigma_{eff})$ being a state funtion depending on the constitutive equation relating the uniaxial stress to the uniaxial strain.

For a material that is describable by the Ramberg–Osgood relation

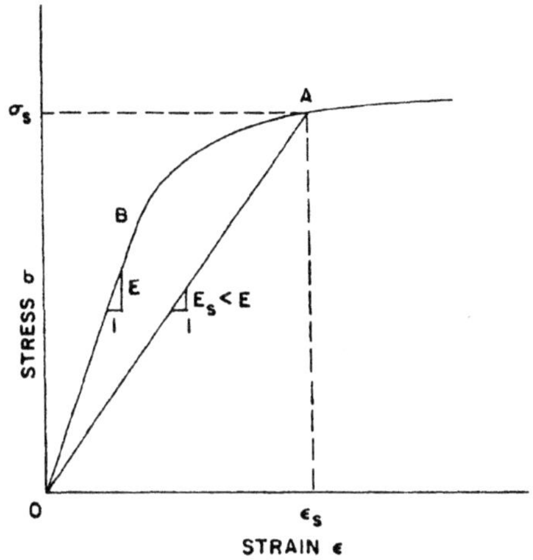

Fig. 5.3: Path of deformation in plasticity.

$$\varepsilon = \begin{cases} \dfrac{\sigma}{E} & \text{, for } \sigma < \sigma_{yd} \\ \dfrac{\sigma}{E} + \dfrac{m\sigma_{yd}}{E}\left[\left(\dfrac{\sigma}{\sigma_{yd}}\right)^n - 1\right] & \text{, for } \sigma \geq \sigma_{yd} \end{cases} \tag{5.9}$$

the function $f(\sigma_{eff})$ is given by

$$f(\sigma_{eff}) = \begin{cases} \dfrac{2mn}{3E\sigma_{yd}}\left(\dfrac{\sigma_{eff}}{\sigma_{yd}}\right)^{n-2} & \text{, for } \dot{\sigma_{eff}} \geq \sigma_{yd} \text{ and } d\dot{\sigma_{eff}} > 0 \\ 0 & \text{, otherwise.} \end{cases} \tag{5.10}$$

Assumed in plasticity is that the uniaxial response in Equations (5.9) coincides with the effective stress σ_{eff} in Equation (5.2) and the effective strain ε_{eff} whose change is given by

$$d\varepsilon_{eff} = \frac{\sqrt{2}}{3}\Big\{(d\varepsilon_{xx} - d\varepsilon_{yy})^2 + (d\varepsilon_{xx} - d\varepsilon_{zz})^2 + (d\varepsilon_{yy} - d\varepsilon_{zz})^2 + \\ + \frac{3}{2}[d\varepsilon_{xy}^2 + d\varepsilon_{xz}^2 + d\varepsilon_{yz}^2]^{1/2}\Big\}. \tag{5.11}$$

The procedure involves the application of an initial load, say p, such that at least one point in the material reaches the yield condition and a corresponding strain field ε_{ij}^p is obtained. Dividing the total load into increments such that each load increment gives rise to an increment strain $d\varepsilon_{ij}^p$ that can be calculated from Equation (5.7). The total strain corresponding to a load increase from p to $p + dp$ is

$$\varepsilon_{ij}^{p+dp} = \varepsilon_{ij}^p + d\varepsilon_{ij}^p. \tag{5.12}$$

The strains can thus be accumulated in accordance with a certain prescribed load path. Equations (5.5), (5.7), (5.8) and (5.10) will thus give the deformation field from which the stresses can be found.

Instead of the Prandtl–Reuss relation in Equation (5.7), Hencky's rule for incremental plastic strain can also be used:

$$d\varepsilon_{ij}^{p} = d\lambda(\sigma_{eff})d\mathscr{S}_{ij}', \quad \text{for} \quad \sigma_{eff} \geq \sigma_{yd} \quad \text{and} \quad d\sigma_{eff} \geq 0 \tag{5.13}$$

which makes use of $d\mathscr{S}_{ij}'$ as in contrast to the total stress deviator $\mathscr{S}_{ij}'$ in Equation (5.7). Computations based on Hencky's rule, however, become more involved.

5.3 Macroplastic deformation and crack growth

Despite the extensive efforts made in the past to gain insight into the behavior of the crack tip region, our understanding of the physics has fallen far behind because of the lack of theoretical guidance, without which no meaningful experiments* can be carried out. This work will, therefore, be confined within the framework of the mathematical theory of plasticity and its interpretation of the state of affairs near the tip of a macrocrack.

Elastic-plastic stress singularity

The earlier efforts [3, 19] on crack tip plasticity were basically centered on determining the stress field singularity, an extension of the idea behind linear elastic fracture mechanics. Complete homogeneity of plastic deformation were invoked [3, 19] such that a so-called plastic stress intensity factor was defined as the coefficient of the singular term, $r^{-\psi}$, where ψ depended on the strain hardening coefficient in the constitutive relation. This result is unrealistic because the macrostress field around the crack tip need not be completely plastic; it can acquire an elastic-plastic character as illustrated in Figure 5.4. A substantial portion of the material around the crack is in the elastic regime and retains the inverse square root of r singular behavior. The units of the coefficient for the $1/\sqrt{r}$ term in the elastic portion and $1/r^{\psi}$ term in the plastic portion are not the same. That is to say, the amplitude of the elastic-plastic stress field becomes nonhomogeneous and is no longer representative by a single intensity factor in terms of stress or strain. By the same token, the J-integral [23] approach does not apply because the value of J would not be single-valued; it would depend on the contour size of integration. As r is varied in Figure 5.4, the portion of elastic and plastic region within the dotted semi-circle would change accordingly. Besides, the J-

* Recent analyses and experiments [20 – 22] have shown that damage ahead of a crack is a highly nonequilibrium/irreversible process that differed both qualitatively and quantitatively from the plasticity theory prediction by a wide margin. The inherent discrepancies cannot be easily reconciled as they lie in the very development of the classical continuum mechanics theories that have intrinsically separated thermal change with mechanical deformation; they apply only for equilibrium states.

integral applies only to the deformation theory of plasticity that assumes path independency which is in contrast to the elastic-plastic deformation of cracks [24 – 28].

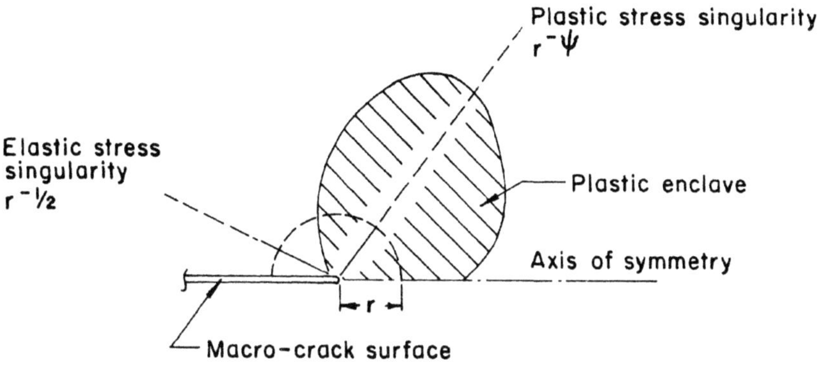

Fig. 5.4: Elastic and plastic singular character of crack.

Strain energy density function

The creation of new surface involves crack initiation and growth that belong to the same physical process. It is only logical that these two events should be governed by one of the same criterion so as to minimize inconsistencies in the outcome. This requirement is not satisfied by many of the proposed criteria, a detailed discussion of which can be found in [27]. Moreover, the restrictions associated with the stress solutions should not be confused with those inherent in the failure criteria or vice versa.

When crack growth is accompanied by permanent deformation of the surrounding material, not all of the energy is available to drive the crack as some of which will be used to distort and dilate the material elements. Unless the portion of the energy used to spread the crack can be separated from the total energy [2], no realistic prediction on crack growth can be made. The inhomogeneous character of the elastic-plastic stresses and/or strains ahead of the crack must also be accounted for in its entity without ambiguity. Generality and consistency are useful guidelines for selecting the appropriate crack growth criterion. Under these considerations, the strain energy density function dW/dV, when postulated appropriately and applied consistently, can satisfy all the necessary requirements mentioned earlier. The strain energy density criterion, in fact, applies to any material; this, of course, includes solids with elastic-plastic behavior [24 – 28]. It is because of the $1/r$ singular character of dW/dV:

$$\frac{dW}{dV} = \frac{S}{r} \tag{5.14}$$

in which S is known as the strain energy density factor and r is the radial distance measured from the location* under investigation.

While S will depend on the theory used in the stress analysis such as elasticity or plasticity, the $1/r$ dependency is independent of the constitutive relation of the material. Figure 5.5 provides a schematic display of the rapid decay of dW/dV as the distance r from the crack tip is increased. The S-factor in Equation (5.14) is the area under the dW/dV versus r curve which depends on the state of stress and strain ahead of the crack. The total energy density dW/dV may be divided into two components:

$$\frac{dW}{dV} = \left(\frac{dW}{dV}\right)_p + \left(\frac{dW}{dV}\right)^* . \tag{5.15}$$

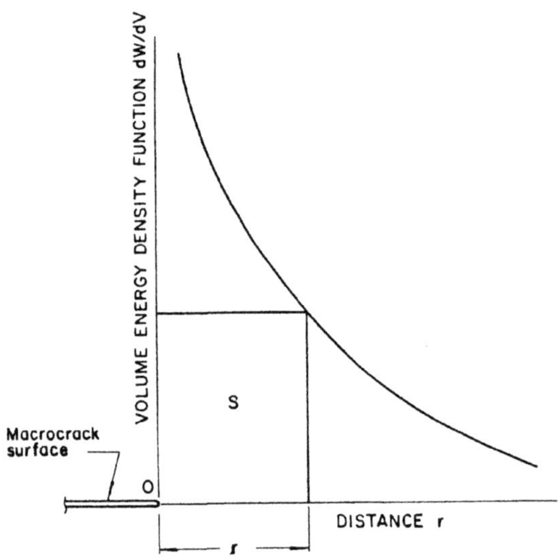

Fig. 5.5: Variations of volume energy density with distance for static and dynamic load of same magnitude.

The portion used in plastic deformation is $(dW/dV)_p$ and the portion available to drive the crack is $(dV/dV)^*$ whose critical value $(dW/dV)_c^*$ is characteristic of the material and can change for each increment of crack growth. Refer to Figure 5.6 for a graphical interpretation of $(dW/dV)_p$ and $(dW/dV)^*$ referred to the uniaxial true stress and true strain curve where dW/dV represents the total area under the curve for a given state p. As p reaches f, dW/dV becomes critical or $(dW/dV)_c$.

* It need not be the crack tip since the strain energy density criterion applies equally well to bodies with no initial defects or cracks [29 – 32].

Incremental crack growth

According to the strain energy density criterion [5], crack initiation occurs when the volume energy density dW/dV in the crack tip element reaches the critical value $(dW/dV)_c$; this constitutes a condition of local failure. If little or no energy is lost in the deformation proces, then crack growth is assumed to follow the relation:

$$\left(\frac{dW}{dV}\right)_c = \frac{S_1}{r_1} = \frac{S_2}{r_2} = \cdots = \frac{S_j}{r_j} = \cdots = \frac{S_c}{r_c} = \text{const.} \tag{5.16}$$

The critical strain energy density factor S_c can be related to the valid ASTM fracture toughness value K_{1c}:

$$S_c = \frac{(1+\nu)(1-2\nu)K_{1c}^2}{2\pi E} \tag{5.17}$$

where E is Young's modulus and ν Poisson's ratio. The condition

$$S_c = r_c\left(\frac{dW}{dV}\right)_c \tag{5.18}$$

corresponds to the termination of subcritical crack growth or the onset of rapid crack extension leading to global instability. In this respect, *the thresholds of crack initiation and propagation are assumed to be determined, respectively, by* $(dW/dV)_c$ *and* S_c with r_c being the ligament of material that initiates instability. For a crack completely engulfed in a uniform stress field, the crack growth segments leading to global instability follow the inequalities:

$$S_1 < S_2 < \cdots < S_j < \cdots < S_c$$
$$r_1 < r_2 < \cdots < r_j < \cdots < r_c. \tag{5.19}$$

Subcritical crack growth corresponds to the incremental failure of crack tip elements; it is basically *a repetitive process of crack initiation.* If the energy supplied to the crack tip decreases with time, crack growth can terminate resulting in crack arrest. Such a situation corresponds to

$$S_1 > S_2 > \cdots > S_j > \cdots > S_0$$
$$r_1 > r_2 > \cdots > r_j > \cdots > r_0 \tag{5.20}$$

where $S_0/r_0 < (dW/dV)_c$.

The conditions described in Equations (5.16), (5.19) and (5.20) also apply to crack growth accompanied by plastic deformation where the energy used to drive the crack $(dW/dV)^*$ must now be distinguished from that dissipated by plastic flow $(dW/dV)_p$ as given in Equation (5.15). Equation (5.18) can be modified as*

* The influence of plasticity on the material ahead of the crack becomes less and less significant as the crack slows down so that the difference between S_0/r_0 and S_0^*/r_0^* is not as important. Refer to [29] for more details on the problem of crack arrest.

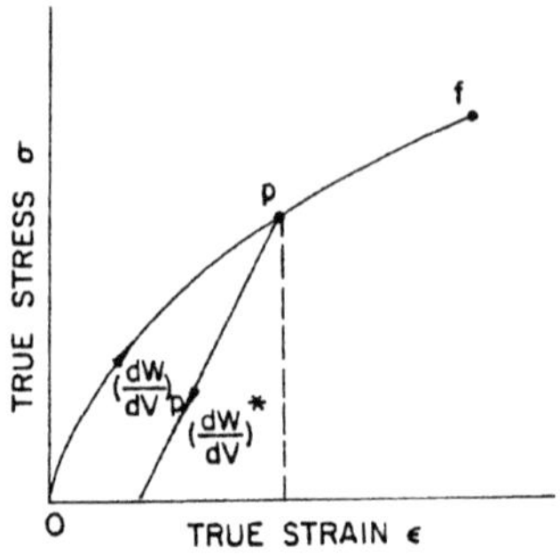

Fig. 5.6: True stress and true strain curve.

$$S_c^* = r_c^* \left(\frac{dW}{dV} \right)_c^* < S_c. \tag{5.21}$$

Each segment of elastic-plastic crack growth must then follow the condition

$$\frac{S_j}{r_j} \to \left(\frac{dW}{dV} \right)_c^*, \; j = 1, 2, \text{etc.} \tag{5.22}$$

Should the material properties, i.e., the stress and strain relation in the crack tip element change for each segment of growth, then $(dW/dV)_c^*$ must also be adjusted for each value of j; $(dW/dV)_c^*$ in Equation (5.22) should then be replaced by $[(dW/dV)_c^*]_j$. The theory of plasticity cannot account for this effect due to the change in local strain rates unless appropriate refinements can be made.

Crack growth resistance

Nonlinearities of the crack growth data are sensitive to changes in specimen size, loading rate and material type. Predictive capability relies on linearization of the data so that interpolation can be applied to forecast results other than those tested. To this end, the strain energy density factor S can again serve a useful purpose. The rate change of S with respect to crack length, say $\Delta S/\Delta a$, is assumed to be a constant during crack growth, i.e.,

$$\frac{\Delta S}{\Delta a} = \text{const.} \tag{5.23}$$

This condition has been shown to hold for a variety of problems [7 – 14] both in static and fatigue loading.

The usefulness of Equation (5.23) is illustrated graphically in Figures 5.7(a) to 5.7(c) inclusive. They represent typical data obtained from cracked metal panels [2, 9, 10]. As the specimen size, characterized by the ratio of the initial volume to surface area V/A, is varied for a given material and loading rate, S can be computed by application of the incremental theory of plasticity described in Section 5.2 for each increment of crack growth. The condition in Equation (5.23) is satisfied for different

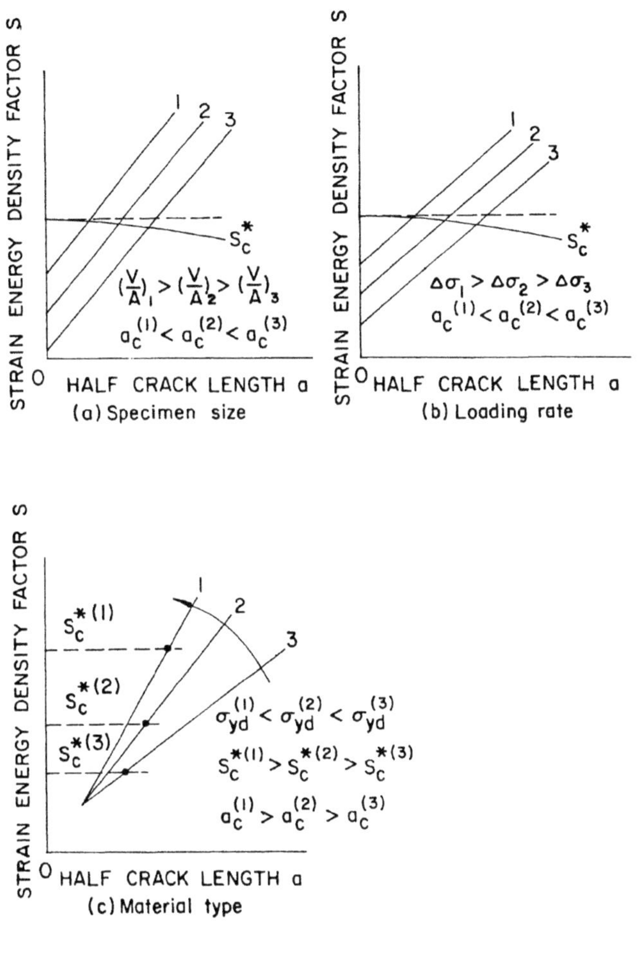

Fig. 5.7: Crack growth scaling of specimen size, loading step and material type.

V/A ratios as illustrated in Figure 5.7(a). Smaller size specimen enhances plastic deformation and subcritical crack growth; this corresponds to the straight line with $(V/A)_3$ where the corresponding distance from 0 to $a_c^{(3)}$ is great than those for the larger specimens with $(V/A)_1$ and $(V/A)_2$. Here, the intersections of the $\Delta S/\Delta a$ = const. line with S_c^* give the critical crack lengths $a_c^{(1)}, \ldots, a_c^{(3)}$ that represent the onset of unstable fracture. Specimens with large volume to surface ratio tend to behave more brittle as expected. A trend between the spacing of the $\Delta S/\Delta a$ lines and V/A may thus be established to yield additional results by extrapolation. The S_c^* value is not strictly a constant because r_c^* tends to increase with decreasing specimen size and/or loading step. This trend is shown in Figures 5.7(a) and 5.7(b) with S_c^* being slightly curved; the amount of deviation from a straight line depends on a particular material. Similarly, the specimen size and material may be fixed while the loading steps or rates may be varied. The results are exhibited in Figure 5.7(b). Large load steps correspond to fast loading rate and lead to unstable fracture with less subcritical crack growth. This behavior has also been observed repeatedly in experiments. The $\Delta S/\Delta a$ = const. lines rotate counterclockwise in the S versus a plot as the fracture toughness* S_c or S_c^* of the material is increased. Figure 5.7(c) shows that the three materials with $S_c^{*(1)}, ..., S_c^{*(3)}$ will interact the straight lines at three different levels and subcritical crack growth increases with increasing S_c^*.

Although the condition of $\Delta S/\Delta a$ = const. (5.23) holds, in general, for all theories and materials, the results displayed in Figures 5.7(a) to 5.7(c) inclusive are valid only for the idealized elastic-plastic material. There are a great number of simplifying assumptions made in the incremental theory of plasticity that cannot be justified on physical grounds. The lines in Figure 5.7(a) would collapse into one if the exchange of surface and volume energy is considered [31, 32] so that thermal fluctuations would be synchronized with mechanical deformation during crack growth. A family of horizontal lines would be obtained in Figure 5.7(b) for a given material subjected to different loading rates. While the fracture toughness parameter S_c or S_c^* is specimen size invariant for a given material, it is loading rate dependent.

5.4 Inhomogeneous deformation: change in local strain rates

One of the most serious limitations of plasticity theory lies in its inability to account for change in local strain rates and strain rate history. This effect cannot be ignored in regions near geometric discontinuities and surfaces or interfaces where the mechanism of energy dissipation is dominated by distortion in contrast to that of dilatation experienced by elements in the interior of the solid**.

Dilatation and distortion

Separation of distortional and dilatational effects is, of course, not possible when nonlinearity is present. This difficulty does not arise in the strain energy density

* Fracture toughness is associated with the sudden release of available energy to trigger global instability. What has been dissipated in plastic deformation prior to this event becomes relevant only in determining the crack driving force. The meaning of fracture toughness for plastically deformed materials is discussed in [30].

** A detailed explanation on the ways with which the dilatational and distortional components of the volume energy density function affect crack growth can be found in [5] (Volume VII).

criterion [5] which automatically determines the proportion of distortion and dilatation from the relative stationary values. Yielding, being dominated by distortion, is assumed to coincide with the relative maximum of dW/dV or $(dW/dV)_{max}$ while fracture coincides with the relative minimum of $(dW/dV)_{min}$. Since there may be many maxima and minima, the sites where yielding and fracture would initiate can thus be located by finding the maximum of $(dW/dV)_{max}$ and $(dW/dV)_{min}$. Such a scheme has been used widely and successfully in engineering problems with or without preexisting defects or cracks.

The theory of plasticity, however, is not free from violating the separation of distortional or dilatational component of energy density from the total energy density dW/dV. A case in point is the von Mises yield criterion which uses only the distortional component of dW/dV. Attempts have been made to modify the J_2'-flow rule by incorporating the effect of dilatation for determining the onset of yielding. Convexity of a yield function and positive definiteness of the plastic work were invoked for generating a class of admissible yield functions in the stress space [33]. A special form for such functions is

$$F(J_1, J_2') = J_2' + \alpha^2(J_1 + \beta)^2 \tag{5.24}$$

in which

$$J_1 = \mathrm{Tr}[\sigma] = 3\dot{\sigma}_1, \quad J_2' = \mathrm{Tr}[\sigma - \overline{\sigma}_1 I]^2 \tag{5.25}$$

with $\overline{\sigma}_1$ being the average of the normal stresses. The dependency of F on J_1 is weighted through α and the translation of the yield surface in the direction of the hydrostatic stress is governed by β. These are empirically determined parameters. Similar forms of Equation (5.24) have also been pointed out in [34]. The addition of J_1 in the yield function, however, is not consistent with the assumption that the uniaxial data coincide with the effective stress and effective strain where the influence of dilatation is also omitted. These deficiencies can lead to inaccurate description of the crack profile in regions where volume change is not negligible; this can occur in a surface layer of material not so far away from the free surface where both the volume and surface energy density play an active role*. For instance, the growth of a thumbnail-shaped crack in a finite thickness plate would adopt a *kink* in the profile near the surface layer. This would occur if the stress and strain solutions derived from either the classical theory of elasticity or plasticity are used in conjunction with a particular failure criterion.

---

* A bump occurred in the elastic crack front stress intensity factor through the plate thickness [35]. This is not a numerical inaccuracy of the finite element method but a fundamental deficiency in all classical continuum mechanics theories by letting the rate change of volume with surface to vanish in the limit, i.e., $dV/dA \rightarrow 0$ as the element size shrinks. It is the decoupling of surface and volume effect that gives rise to the phenomenon of an irregular crack profile. This does not occur in a theory that considers the exchange of surface and volume energy [31].

Correction for nonhomogeneity

An extensive study was made [10] to remedy the inadequacy of plasticity when applied to regions near a free surface. The problem of crack growth in a finite thickness plate was depicted because the results would not be obscured by assumptions invoked in the two dimensional theories. Moreover, crack profile predictions can be directly verified by experiments. Corrections on the nonhomogeneous character of deformation needs to be introduced. This is accomplished by assuming that the elastic-plastic material properties can vary in the thickness direction of the plate. Different constitutive relations are assigned to different locations. The yield strength in elements near the free surface would be lower than that in the interior; a gradient of yield strength and, hence, fracture toughness is modelled. Such an approach resulted not only in realistic crack profiles but also predicted experimental K_{1c} values from uniaxial data with accuracy [10]. Because of the important nature of this correction for nonhomogeneity in the elastic-plastic deformation field, an example will be provided in the section to follow.

Elastic-plastic crack growth

Nonhomogeneity in material due to change in local strain rates is location and material specific. If ε_{ij} denote the strain components local to the crack tip and $\varepsilon(t)$ the strain on the specimen boundary, then the ratio

$$\frac{\Delta\varepsilon_{ij}/\Delta t}{\Delta\varepsilon/\Delta t} = G(E, \sigma_{yd}, m, n, r, \theta, t) \tag{5.26}$$

depends on the near tip polar coordinates (r, θ) and the time t in addition to the material parameters E, σ_{yd}, m and n in Equation (5.9) that may also change from element to element. The specifics can be best illustrated in connection with the elastic-plastic deformation of a plate with a central crack.

A rectangular plate 10 in. wide and 20 in. long contains a crack 2 in. in length and is subjected to a constant rate of stress $\dot{\sigma} = 3 \times 10^5$ ksi/sec as given in Figure 5.8(a). Only one-quarter of the cracked plate needs to be discretized by employing 26 isoparametric finite elements with 161 nodes while a total of sixty-four (64) Gaussian points is used in each element. This is shown in Figure 5.8(b) in which Figure 5.8(c) gives an enlarged view of the crack tip elements. The $1/9$ and $4/9$ sliding node technique [36] is used to ensure the $1/r$ singular character for the strain energy density field.

Material data bank

Corrections for inhomogeneity in the local strains are made possible by a material data bank that contains uniaxial stress and strain curves obtained experimentally under different strain rates. Typical variations of the uniaxial stress and strain response with strain rate $\dot{\varepsilon}$ are exhibited by the curves in Figure 5.9 which covers five (5) orders of

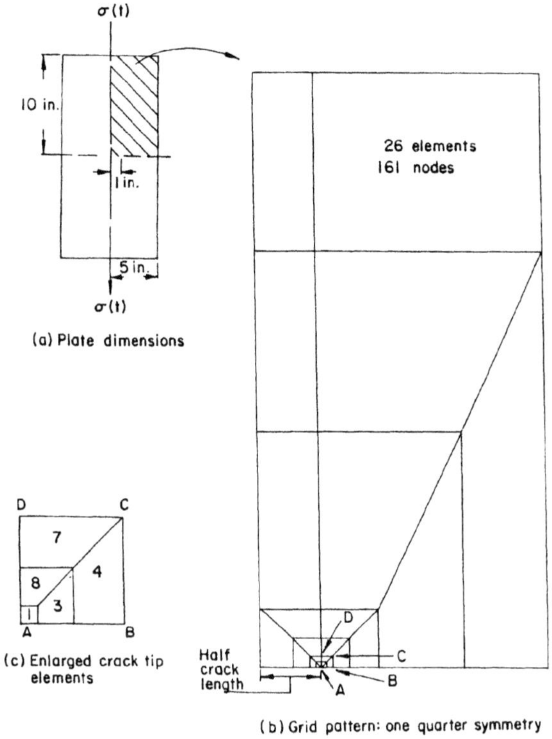

Fig. 5.8: A center-cracked plate subjected to constant rate of stress $\dot{\sigma} = 3 \times 10^5$ ksi/sec.

magnitude of $\dot{\varepsilon}$ from 10^{-3} to 10^2 sec^{-1}. The corresponding values of the material parameters in the Ramberg–Osgood relation in Equation (5.9) can be found in Table 5.1. Only the effective strain ε_{eff} and its change with respect to time $\dot{\varepsilon}_{eff}$ are needed to locate a new state in the material data bank. This is because the data bank is constructed in such a way that ε_{eff} and $\dot{\varepsilon}_{eff}$ can uniquely identify each set of the material parameters in Table 5.1. Such a procedure can be repeated for each load step until a complete stress and strain curve is developed which will be different for each element in the vicinity of the crack.

Incremental loading

Let the cracked plate in Figure 5.8(a) be loaded in intervals of $\Delta t = 10\,\mu$sec at the rate of $\dot{\sigma} = 3 \times 10^5$ ksi/sec. The kth load level is then given by $\sigma^{(k)} = \dot{\sigma} k \Delta t$ and the corresponding incremental strain is related to the incremental stress as

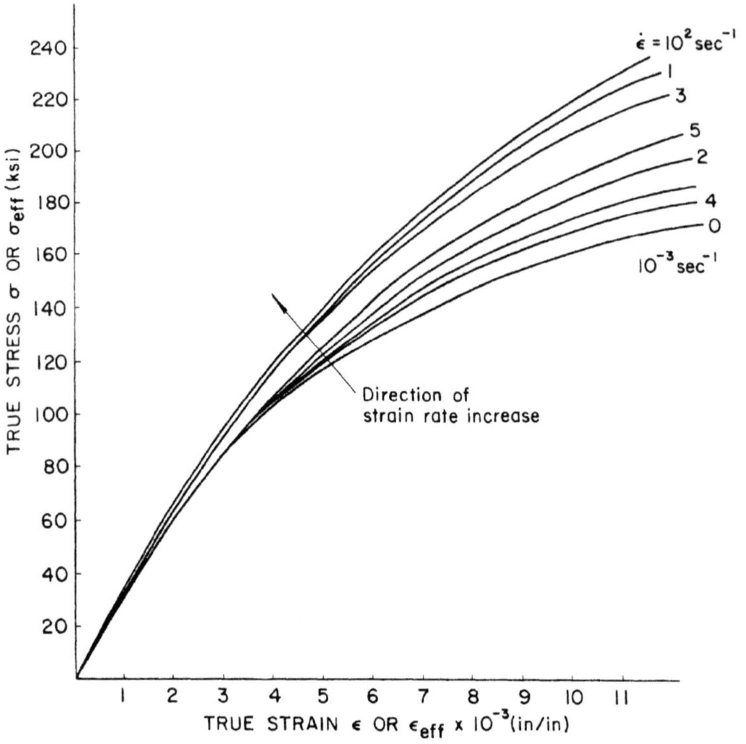

Fig. 5.9: Uniaxial stress-strain curve with different rate of strain.

Table. 5.1: Material parameters in the Ramberg – Osgood relation.

Curve No.	Young's Modulus $E \times 10^3$ (ksi)	Yielding Stress σ_{yd} (ksi)	Hardening Coefficient and Exponent	
			m	n
0	30	60	0.02	5
1	32	70	0.05	3
2	30	60	0.03	4
3	32	70	0.02	4
4	30	60	0.04	4
5	30	60	0.02	4

$$de_{ij}^{(k)} = \begin{cases} \frac{1+\nu}{E}\, d\mathscr{S}_{ij}^{(k)} + 3f\,(\sigma_{eff}^{(k)})\mathscr{S}_{ij}^{(k)}\, d\sigma_{eff}^{(k)} & ; \text{for } d\sigma_{eff}^{(k)} > 0 \text{ and } \sigma_{eff}^{\prime(k)} \geq \sigma_{yd}^{(k)} \\ 0 & ; \text{otherwise} \end{cases} \tag{5.27}$$

in which

$$\sigma_{eff}^{\prime(k)} = \sqrt{\frac{3}{2} \mathcal{S}_{ij}^{\prime(k)} \mathcal{S}_{ij}^{\prime(k)}} \tag{5.28}$$

stands for the effective stress deviator at the kth step. The Ramberg – Osgood relation for the uniaxial state of stress and strain must also be modified:

$$\varepsilon^{(k)} = \begin{cases} \dfrac{\sigma^{(k)}}{E} & ; \text{for } \sigma^{(k)} \leq \sigma_{yd}^{(k)} \\ \dfrac{\sigma^{(k)}}{E} + \dfrac{m^{(k)}\sigma_{yd}^{(k)}}{E}\left[\left(\dfrac{\sigma^{(k)}}{\sigma_{yd}^{(k)}}\right)^{n^{(k)}} - 1\right] & ; \text{for } \sigma^{(k)} > \sigma_{yd}^{(k)}. \end{cases} \tag{5.29}$$

It follows that

$$f\,(\sigma_{eff}^{(k)}) = \begin{cases} \dfrac{2m^{(k)}n^{(k)}}{2E\sigma_{yd}^{(k)}}\left[\dfrac{\sigma_{eff}^{(k)}}{\sigma_{yd}^{(k)}}\right]^{n^{(k)}-2} & ; \text{for } \sigma_{eff}^{\prime(k)} \geq \sigma_{yd}^{(k)} \\ 0 & ; \text{otherwise.} \end{cases} \tag{5.30}$$

A prime has been used to denote measurement with reference to the yield surface at the kth step. For instance,

$$\mathcal{S}_{ij}^{\prime(k)} = \mathcal{S}_{ij}^{(k)} - C_{ij}^{(k)}. \tag{5.31}$$

The incremental change of the center of the yield surface in the kth step is assumed to be in the same direction of the current total stress deviator $\mathcal{S}_{ij}^{\prime(k)}$ whose magnitude is assumed to be proportional to the projection of the total stress deviator onto the local normal of the current yield surface with $\sigma_{yd}^{(k)}$;

$$dC_{ij}^{(k)} = \begin{cases} \dfrac{3\lambda \mathcal{S}_{kl}^{\prime(k)}\, d\mathcal{S}_{kl}^{\prime(k)}\, \mathcal{S}_{ij}^{\prime(k)}}{2(\sigma_{eff}^{\prime(k)})^2} & ; \text{for } d\sigma_{eff}^{(k)} \geq 0 \text{ and } \sigma_{eff}^{(k)} \geq \sigma_{yd}^{(k)} \\ 0 & ; \text{otherwise.} \end{cases} \tag{5.32}$$

For simplicity, λ is assumed to be independent of strain rate change and a value of 0.5 will be used in the subsequent numerical calculations. In general, it depends on hardening characteristics of the material for a given load history.

The state variables in Equations (5.27) – (5.32) with the superscript k will be calculated for each time step as corrections are made on the material parameters $\sigma_{yd}^{(k)}$, $m_{yd}^{(k)}$, etc., in accordance with the local strain rate. Figure 5.10 summarizes the different segments of the effective stress and effective strain at the point A directly above the crack tip at a distance 3×10^{-2} in. At $t = 10\ \mu$sec, ε_{eff} and $\dot{\varepsilon}_{eff}$ are found to be, respectively, 2.8944×10^{-3} in/in and 38 sec^{-1}. A search is then made in the material data bank to locate a particular curve with same ε_{eff} and $\dot{\varepsilon}_{eff}$; this determines the segment of the $\sigma_{eff} - \varepsilon_{eff}$ curve between $t = 10$ and 20 μsec as indicated in Figure 5.10. The same procedure is repeated up to $t = 40\ \mu$sec. A significant difference is found between the original curve labelled 'base material' that would have remained unchanged if the constitutive relation for the local element were not modified. The nonhomogeneous response of elements located at different parts of the cracked plate up to $t = 50\,\mu$sec is exhibited in Figure 5.11 for element 1 nearest to the crack tip; element 8 on top of the crack front region; and element 26 far away from the crack. As expected, the behavior of element 26, which is not influenced by the crack, agreed closely with that of the base material. Element 1 acquired a much higher strain rate and deviated the most from the dotted curve. The change in the slope of the stress and strain curve is significant and can correspond to several orders of magnitude difference in the strain rate.

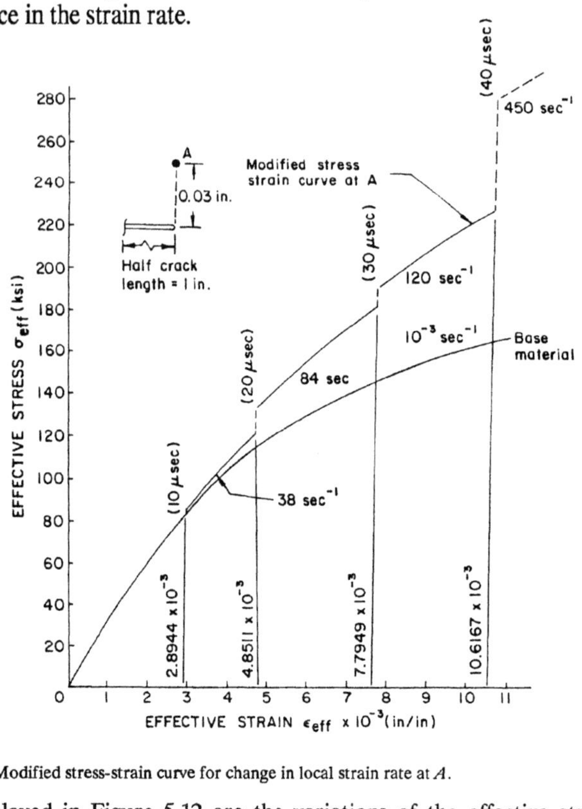

Fig. 5.10: Modified stress-strain curve for change in local strain rate at A.

Displayed in Figure 5.12 are the variations of the effective stress with effective strain for the point B that is located on the path of prospective path of crack growth; again, it deviated significantly from the base material. If parallel unloading is assumed, the critical available energy density $(\mathrm{d}W/\mathrm{d}V)^* = 0.67629$ ksi can be determined. The occurrence of crack growth will depend on whether $(\mathrm{d}W/\mathrm{d}V)^*$ surpasses the critical value $(\mathrm{d}W/\mathrm{d}V)_c^*$ for the material or not. Figure 5.13 gives the typical trade off relation between the yield strength and fracture toughness for an ordinary structural steel. For $\sigma_{yd} = 144$ ksi, the $(\mathrm{d}W/\mathrm{d}V)_c^*$ value is slightly below $(\mathrm{d}W/\mathrm{d}V)^*$ at point B; this implies crack growth. The situation is summarized in Figure 5.14 where the intersection of

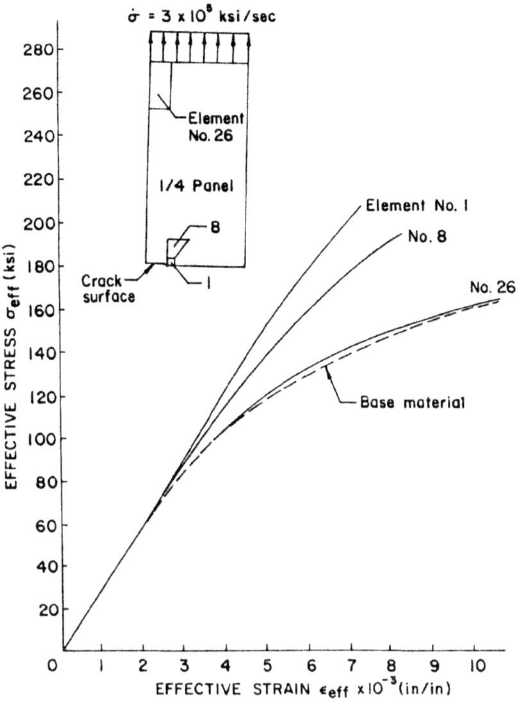

Fig. 5.11: Nonhomogeneous response of material elements for time up to 50 μsec.

$(\mathrm{d}W/\mathrm{d}V)_c^* = 0.66844$ ksi for the material under consideration with the $\mathrm{d}W/\mathrm{d}V$ versus r curve gives an increment of crack growth 5.582×10^{-3} in. at $t = 50$ μsec. Had plasticity been used without modification, the amount would be 1.137×10^{-3} in. It differs by a factor of almost five (5) and represents an underestimate if the elevation of local strain rate is not accounted for. This difference cannot be overlooked and shows that the constitutive relation for material elements at large does not hold in regions near geometric discontinuities.

5.5 Limitations of plasticity and possible alternatives

The theory of plasticity is constructed primarily from the concept of yield function in the stress space and is restricted to equilibrium states as in all classical continuum mechanics theories. Conceptual difficulties, therefore, arise in attempting to evaluate the constitutive coefficients from uniaxial test data which correspond to thermal/mechanical, nonequilibrium/irreversible states. Such fundamental inconsistencies cannot be swept under the rug because they lead to serious misrepresentations when applied to situations that violate the basic assumptions of the theory. Plasticity works well when applied as in limit analysis for predicting the collapse load of structures but becomes highly questionable when employed to analyze

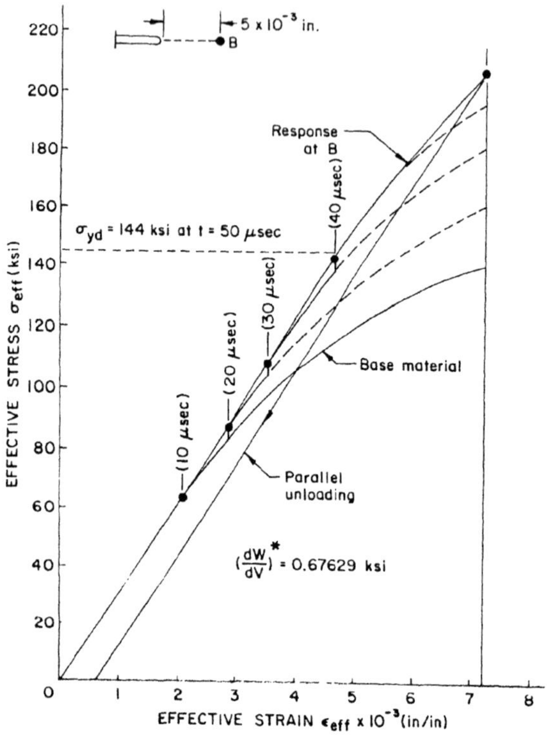

Fig. 5.12: Effective stress and effective strain response at point *B*.

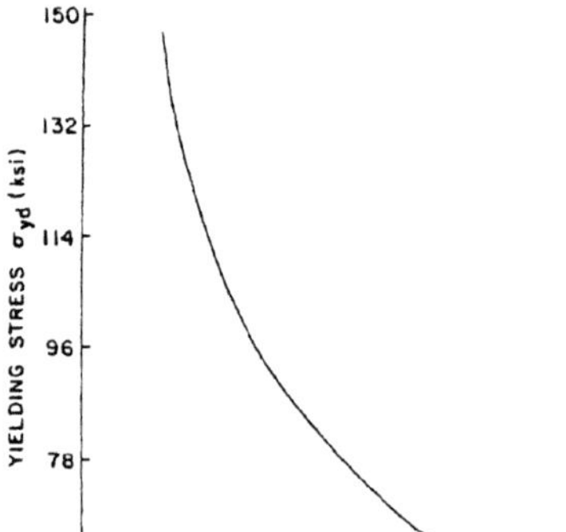

Fig. 5.13: Variations of yield strength with fracture toughness.

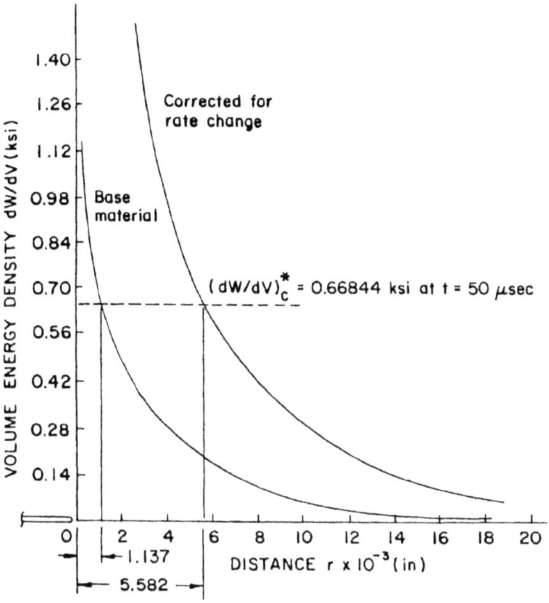

Fig. 5.14: Crack growth at $t = 50$ μsec.

details near geometric discontinuities, such as cracks where the local strain rates can be many times higher than those at remote distances. Unless careful thoughts are made to guard against hidden subleties, modifications to plasticity should not be made without considering the underlying philosophy of the basic approach. The mathematical admittance of J_1 into the yield function is not sufficient and may even introduce inconsistencies. This would indeed be the case if the coincidence of uniaxial data with the effective stress and effective strain response is still retained. *Consistency* requires that the yield function should admit only J_2 as in the von Mises criterion and not J_1. The dilemma arises from the establishment of a one-to-one correspondence between data from the uniaxial tests and multiaxial stress and/or strain states, a condition that is well-recognized but seldom brought to light for the lack of an alternative.

Even though the discussions in Section 5.4 offered a way to rectify the change in local strain rates, there remains the issue of irreversible energy dissipation which is treated empirically in plasticity; the path of unloading, for example, can only be determined empirically and not by theory. Thermodynamics and plasticity have traditionally been divorced from any reconciliations in view of their separate identities. This is why temperature change and elastic-plastic deformation are not synchronized. The cooling portion* of the curve in uniaxial tests cannot be predicted by plasticity

* The temperature near the crack tip was below ambient for more than three (3) minutes in the compact tension specimen made of 1020 steel [21], a result that is not intuitively obvious.

which assumes heating only. What cannot be ignored is the qualitative feature of cooling/heating and the mutual interaction of temperature and deformation; thermal and mechanical equilibrium or nonequilibrium must be addressed simultaneously. Thermal and mechanical recoverability of metals and polymers or rubbers can be quite different. Rubber may recover well mechanically but poorly in repeating the thermal history. It is only in recent years that a general means of coupling thermal and mechanical effects has been found [39, 40]. This is accomplished by not shrinking the continuum element to zero in the limit. The change of volume with surface area, dV/dA, no longer vanishes, which represents a fundamental departure from classical continuum mechanics. Nonequilibrium/irreversible processes can be analyzed in a way that is free from ambiguities so that the temperature Θ and strain e_0 are directly related [39, 40]:

$$\frac{\Delta\Theta}{\Theta} = -\lambda_0\left(\frac{\Delta V}{\Delta A}\right)_0 \frac{\Delta e_0}{\Delta\mathscr{D}/\Delta e_0}. \tag{5.33}$$

The subscript 0 refers to an isoenergy element which has the unique feature that the same energy is transmitted in all directions making e_0 the isostrain. Nonequilibrium uniaxial data can thus be correlated to multiaxial strain states without loss in generality. In Equation (5.33), $\mathscr{D}$ stands for the energy dissipation that is derived analytically and λ_0 determines the strain state under consideration. It is the term $(\Delta V/\Delta A)_0$ neglected in continuum mechanics theories that reflects the cooling/heating behavior. The aforementioned approach has now been well established and has solved numerous engineering problems by application of the PEDDA code [41] which differs fundamentally from DEPA [36] using the theory of plasticity. Correction for mechanical strain rates alone is not sufficient in any theory that attempts to consistently account for energy dissipation and irreversibility, thermal changes have to be included as an inherent part of mechanical deformation.

References

[1] SIH, G.C., *Fracture Mechanics in Engineering Application,* Edited by G.C. Sih and S.R. Valluri, Sijthoff and Noordhoff (now Kluwer Academic Publishers), The Netherlands, 1979, pp. 3–29.

[2] SIH, G.C. and MADENCI, E., *Journal of Engin. Fract. Mech.,* Vo. 18, No. 6, 1983, pp. 1159–1171.

[3] HUTCHINSON, J.W., *Journal Mech. Phys. Solids,* Vol. 16, 1968, pp. 337–342.

[4] HILTON, P.D. and HUTCHINSON, J.W., *Journal of Engin. Fract. Mech.,* Vol. 3, 1971, pp. 435–451.

[5] SIH, G.C., *Mechanics of Fracture,* Introductory Chapters of Vols. I to VII, Martinus Nijhoff (Kluwer Academic Publishers), The Netherlands, 1972–1987.

[6] IVANOVA, V.S., MASLOV, L.J. and BURBA, V.I., *Journal of Theoretical and Applied Fracture Mechanics,* Vol. 2, No. 3, 1984, pp. 223–227.

[7] GDOUTOS, E.E., *Journal of Theoretical and Applied Fracture Mechanics,* Vol. 1, No. 2, 1984, pp. 139–144.

[8] CARPINTERI, A. and SIH, G.C., *Journal of Theoretical and Applied Fracture Mechanics,* Vol. 1, No. 2, 1984, pp. 145–159.

[9] SIH, G.C. and TZOU, D.Y., *Modelling Problems in Crack Tip Mechanics,* edited by J.T. Pindera, Martinus Nijhoff (Kluwer Academics Publishers), The Netherlands, 1984, pp. 155-169.
[10] SIH, G.C. and CHEN, C., *Journal of Theoretical and Applied Fracture Mechanics,* Vol. 3, No. 2, 1985, pp. 125-139.
[11] SIH, G.C. and MOYER, E.T., Jr., *Journal of Engin. Fract. Mech.,* Vol. 17, 1983, pp. 269-280.
[12] MOYER, E.T., Jr. and SIH, G.C., *Journal of Engin. Fract. Mech.,* Vol. 19, No. 4, 1984, pp. 643-652.
[13] SIH, G.C. and CHAO, C.K., *Journal of Theoretical and Applied Fracture Mechanics,* Vol. 1, No. 3, 1984, pp. 239-247.
[14] SIH, G.C. and CHAO, C.K., *Journal of Theoretical and Applied Fracture Mechanics,* Vol. 2, No. 1, 1984, pp. 67-74.
[15] HILL, R., *Mathematical Theory of Plasticity,* Oxford University Press, 1950.
[16] PRAGER, W., *Introduction to Mechanics of Continua,* Dover Press, New York, 1961.
[17] BRIDGMAN, P.W., *Studies in Large Plastic Flow and Fracture with Special Emphasis in the Effects of Hydrostatic Pressure,* Mc. Graw-Hill, New York, 1952.
[18] REUSS, A., *Z. Ang. Math. Mech.,* Vol. 10, 1930, p. 266.
[19] RICE, J.R. *et al., Journal Mech. Phys. Solids,* Vol. 16, 1968, pp. 1-12.
[20] SIH, G.C. and TZOU, D.Y., *Journal of Theoretical and Applied Fracture Mechanics,* Vol. 6, No. 2, 1986, pp. 103-111.
[21] SIH, G.C., TZOU, D.Y. and MICHOPOULOS, J.G., *Journal of Theoretical and Applied Fracture Mechanics,* Vol. 7, No. 2, 1987, pp. 79-87.
[22] TZOU, D.Y. and SIH, G.C., *Journal of Theoretical and Applied Fracture Mechanics,* Vol. 10, No. 1, 1988.
[23] RICE, J.R., *Journal of Appl. Mech.,* Vol. 35, 1968, pp. 379-386.
[24] SIH, G.C., *Prospects of Fracture Mechanics,* edited by G.C. Sih, H.C. van Elst and D. Broek, Noordhoff International Publishing (now Kluwer Academic Publishers), The Netherlands, 1974, pp. 613-621.
[25] SIH, G.C., *Fracture Mechanics and Technology,* edited by G.C. Sih and C.L. Chow, Sijhoff and Noordhoff (now Kluwer Academic Publishers), The Netherlands, 1977, pp. 967-784.
[26] SIH, G.C., *Modelling Problems in Crack Tip Mechanics,* edited by J.T. Pindera, Martinus Nijhoff (Kluwer Academic Publishers), 1983, pp. 65-90.
[27] SIH, G.C., *Fracture Mechanics Methodology,* edited by G.C. Sih and L. Faria, Martinus Nijhoff (Kluwer Academic Publishers), The Netherlands, 1984, pp. 35-101.
[28] SIH, G.C. and MADENCI, E., *Journal of Engin. Fract. Mech.,* Vol. 78, 1983, pp. 667-677.
[29] SIH, G.C. and TZOU, D.Y., *Journal of Theoretical and Applied Mechanics,* Vol. 6, No. 1,,1986, pp. 29-37.
[30] SIH, G.C., ASTM Technical Publication STP 605, 1976, pp. 3-15.
[31] SIH, G.C., *Journal of Theoretical and Applied Fracture Mechanics,* Vol. 4, No. 3, 1985, pp. 157-173.
[32] TZOU, D.Y. and SIH, G.C., *Journal of Theoretical and Applied Fracture Mechanics,* Vol. 10, No. 1, 1988.
[33] YANG, W.H., *Journal of Appl. Mech.,* Vol. 47, 1980, pp. 297-300.
[34] GDOUTOS, E.E. and SIH, G.C., Institute of Fracture and Solid Mechanics, Technical Report IFSM-84-128, Lehigh University, 1984.
[35] MOYER, E.T., Jr. and LIEBOWITZ, H., *Application of Fracture Mechanics to Materials and Structures,* edited by G.C. Sih, E. Sommer and W. Dahl, Martinus Nijhoff (Kluwer Academic

Publishers), The Netherlands, 1984, pp. 595–606.

[36] SIH, G.C. and TZOU, D.Y., *Dynamic Elastic-Plastic Analysis* (*DEPA*), October, 1986.

[37] SIH, G.C. and TZOU, D.Y., *Journal of Theoretical and Applied Fracture Mechanics,* Vol. 7, No. 1, 1987, pp. 23–30.

[38] SIH, G.C., LIEU, F.L. and CHAO, C.K., *Journal of Theoretical and Applied Fracture Mechanics,* Vol. 7, No. 2, 1987, pp. 67–78.

[39] SIH, G.C., Thermal/Mechanical Interaction Associated with the Micromechanisms of Material Behavior, Institute of Fracture and Solid Mechanics Monograph, Lehigh University, February, 1987.

[40] SIH, G.C., *Journal of Theoretical and Applied Fracture Mechanics,* Vol. 9, No. 3, 1988.

[41] *Plane Energy Density Damage Analysis* (*PEDDA*), Institute of Fracture and Solid Mechanics, Lehigh University, U.S. Library of Congress No. 86-82825, October, 1986.

T.C.T. Ting

6

Nonexistence of higher order discontinuities across elastic/plastic boundary in elastic-plastic wave propagation

6.1 Introduction

The basic theory of elastic-plastic wave propagation was formulated independently in the 1940s during the World War II by Karman [1, 2], Taylor [3, 4] and Rakmatulin [5, 6], and is often referred to as the Karman–Taylor–Rakhmatulin theory. A little known fact is that Donnell [7] had in fact studied the problem earlier in 1930. The theory was for rate-independent materials and was found later to be inadequate to describe the response of real metals under a high strain-rate loading. However, for rate-insensitive materials or for wave propagation in metals in which the strain-rate changes are not too drastic, the theory provides reasonable descriptions of wave propagation in an elastic-plastic material. An extended treatment on the subject can be found in the book by Cristescu [8].

One of the difficulties in finding the solution to problems of elastic-plastic wave propagation is the determination of the elastic-plastic boundary between the elastic region and the plastic region [9 – 16]. Exact solutions for the boundary are difficult to obtain except for special cases and approximate methods have to be employed. Inherited with the elastic-plastic boundary is the discontinuity in the spatial and time derivatives of the stress σ and the velocity v across the boundary. Karman, Bohnenblust and Hyers [17] and Lee [13] were among the first to discuss the discontinuities and obtain the relation

$$\frac{\sigma_t^p}{\sigma_t^e} = \frac{\dfrac{1}{c_e^2} - \dfrac{1}{\lambda^2}}{\dfrac{1}{c_p^2} - \dfrac{1}{\lambda^2}}, \tag{6.1}$$

in which λ is the speed of the elastic-plastic boundary, c is the characteristic wave speed, σ_t is the first order derivative of stress σ with respect to the time t and the superscripts or the subscripts e and p denote, respectively, the elastic and plastic region. Equation (6.1) is useful in determining the range of λ depending on whether the boundary is a loading or an unloading boundary. It can also be used to determine σ_t on the other side of the boundary when λ and σ_t are known on one side of the boundary.

Equation (6.1) is not applicable when $\sigma_t^e = \sigma_t^p = 0$. A modified equation was derived by Clifton and Ting [18] as

$$\frac{\sigma_{tt}^p}{\sigma_{tt}^e} = \frac{\dfrac{1}{c_e^2} - \dfrac{1}{\lambda^2}}{\dfrac{1}{c_p^2} - \dfrac{1}{\lambda^2}}. \tag{6.2}$$

This is the same as Equation (6.1) except that the first order derivatives σ_t^e and σ_t^p are replaced by the second order derivatives σ_{tt}^e and σ_{tt}^p, respectively. Likewise, if the first $(n-1)$th order derivatives of σ with respect to t vanish, Equation (6.1) is modified by replacing σ_t^e and σ_t^p by $\partial^n\sigma^e/\partial t^n$ and $\partial^n\sigma^p/\partial t^n$, respectively. Recently, in a series of papers, Yu, Wang and Zhu [19 – 21] have shown that a discontinuity of order higher than two cannot occur on a finite segment of the elastic-plastic boundary Γ unless $\lambda = c_e$ or c_p, i.e., unless Γ coincides with the characteristics of the elastic or plastic region. This is a remarkable result which is not obvious by inspection of Equations (6.1), (6.2) and their modified version for higher orders.

Although a higher order discontinuity cannot occur on a finite segment of Γ, it is possible to have a higher order discontinuity at an isolated point M on Γ. An example in which σ_t^e and σ_t^p are both nonzero on Γ except at the isolated point M where both σ_t^e and σ_t^p vanish was given by Ting [22]. In this case, neither Equation (6.1) nor (6.2) is applicable at M. The new relations for this case are (see Equation (13) and (18) of [22])

$$\frac{\sigma_{tt}^p + \lambda\sigma_{xt}^p}{\sigma_{tt}^e + \lambda\sigma_{xt}^e} = \frac{\dfrac{1}{c_e^2} - \dfrac{1}{\lambda^2}}{\dfrac{1}{c_p^2} - \dfrac{1}{\lambda^2}}, \tag{6.3a}$$

$$\left(1 + \frac{\lambda^2}{c_e^2}\right)\sigma_{tt}^e + 2\lambda\sigma_{xt}^e = \left(1 + \frac{\lambda^2}{c_p^2}\right)\sigma_{tt}^p + 2\lambda\sigma_{xt}^p. \tag{6.3b}$$

In the papers by Yu, Wang and Zhu, they also extended Equations (6.3) to the cases in which M has a higher order discontinuity.

We will analyze the discontinuities across an elastic-plastic boundary systematically from the first order discontinuities to nth order discontinuities. The order of discontinuity at an isolated point M on the elastic-plastic boundary Γ is equal to or higher than that on Γ. Many results obtained in [19 – 21] are recovered here; some with fewer restrictions. For instance, Theorem 3 of [21] states that if $\sigma_{tt}^e = \sigma_{tt}^p = 0$ on Γ, all second order derivatives of σ and v vanish. As we will see from Theorem 6.5.2 in the following, one need only assume that $\sigma_{tt}^e = \sigma_{tt}^p$. Then all second order derivatives including σ_{tt}^e and σ_{tt}^p vanish. We also obtain a few new results. For instance, we show that on a loading boundary, a discontinuity of order higher than one cannot occur if the yield stress σ_Y which may depend on the position x is such that $\partial^2\sigma_Y/\partial x^2 \leq 0$. This includes the special case in which the material is loaded to a plastic state for the first time and $\partial^2\sigma_Y/\partial x^2 = 0$. Thus, as pointed out by Yu, Wang and Zhu, if the material has never been loaded into a plastic state, a discontinuity of order higher than one cannot occur on a loading boundary unless the boundary coincides with the characteristics of the elastic or plastic region.

6.2 Basic equations for longitudinal waves in a thin rod

We consider longitudinal waves in a thin rod of elastic-plastic materials. The material is assumed to be rate-independent. Let σ and ε be the stress and strain, respectively. For a monotonically increasing load from the virgin state $\sigma = \varepsilon = 0$, the stress-strain law is given by

$$\sigma = E\varepsilon, \quad \text{if} \quad \sigma \leq \sigma_0, \quad \text{(elastic)}, \tag{6.4a}$$

$$\sigma = f(\varepsilon), \quad \text{if} \quad \sigma > \sigma_0, \quad \text{(plastic)}, \tag{6.4b}$$

in which E is Young's modulus, σ_0 is the initial yield stress and f is a prescribed function of ε. Moreover, if $\sigma_0 = E\varepsilon_0$, $\sigma_0 = f(\varepsilon_0)$ so that the stress-strain curve is continuous at $\sigma = \sigma_0$. When $\sigma > \sigma_0$, the material is in a plastic region and an unloading from, and a subsequent reloading to the same stress level, is assumed to be elastic. Thus, if an unloading takes place at the stress $\sigma = \sigma_Y > \sigma_0$, we have

$$\sigma - \sigma_Y = E(\varepsilon - \varepsilon_Y), \quad \sigma_Y = f(\varepsilon_Y), \quad \text{(elastic)}, \tag{6.5}$$

for $\sigma \leq \sigma_Y$ in which σ_Y is the new yield stress. Reloading to $\sigma = \sigma_Y$ follows Equation (6.5) and further loading to $\sigma > \sigma_Y$ follows Equation (6.4b). For the purposes of the present paper, there is no need to consider the possibility of a reverse plastic loading. We therefore assume that both σ and ε are positive.

The wave speed c is related to the slope of the stress-strain curve by

$$\rho c^2 = \frac{d\sigma}{d\varepsilon}, \tag{6.6}$$

where ρ is the mass density. In view of Equations (6.4), we have

$$\begin{aligned} \rho c_e^2 &= E\varepsilon, && \text{in the elastic region,} \\ \rho c_p^2 &= \frac{d}{d\varepsilon} f(\varepsilon), && \text{in the plastic region.} \end{aligned} \tag{6.7}$$

The subscripts (and later on the superscripts) e and p stand for elastic and plastic regions, respectively. We see that the elastic wave speed c_e is a constant while the plastic wave speed c_p depends on the strain ε or, by Equation (6.4b), on the stress σ. We will assume that the stress-strain curve is concave to the strain axis so that c_p is a decreasing function of σ. Hence $c_e > c_p$ except possibly at $\sigma = \sigma_0$ where $c_e = c_p$ if $df/d\varepsilon = E$ at $\varepsilon = \varepsilon_0$. Otherwise $c_e \neq c_p$.

Let v be the particle velocity in the x-direction which is taken along the axis of the rod. The equation of motion can be written as

$$\sigma_x = \rho v_t, \tag{6.8}$$

where t is the time and the subscripts x and t denote partial differentiation with respect to these variables. The continuity of the displacement and its partial derivatives lead to the relation

$$v_x = \varepsilon_t,$$

or, regarding ε as a function of σ and using Equation (6.6),

$$v_x = \frac{\sigma_t}{\rho c^2}. \tag{6.9}$$

Equations (6.8) and (6.9) form a system of first-order hyperbolic differential equations for σ and v. In some instances it may be more convenient to write Equations (6.8) and (6.9) in matrix form as

$$w_x - N w_t = 0, \tag{6.10}$$

in which the 2×2 matrix N and the 2×1 matrix w are given by

$$N = \begin{bmatrix} 0 & \frac{1}{\rho c^2} \\ \rho & 0 \end{bmatrix}, \quad w = \begin{bmatrix} v \\ \sigma \end{bmatrix}. \tag{6.11}$$

We will use either Equations (6.8) and (6.9), or (6.10) to analyze the discontinuities across an elastic-plastic boundary which separates an elastic region from a plastic region.

6.3 Discontinuities of first order

In the (x, t) plane let Γ be a boundary between an elastic region and a plastic region. If the equation for Γ is given by

$$x = \hat{x}(t), \tag{6.12}$$

the speed λ of the elastic-plastic boundary is

$$\lambda = \frac{d\hat{x}}{dt} = \hat{x}_t, \tag{6.13}$$

which is a function of t. *We assume that λ is finite and non-zero, i.e.* $0 < \lambda < \infty$, and is continuously differentiable for as many orders as we require. We also assume that *σ and v are continuous across Γ.* If the first order derivatives of σ and v at a point M on Γ are not all continuous, M is said to have a first-order discontinuity. Likewise, if the first $(n - 1)$th order derivatives are continuous but the nth order is not, M is said to have an nth-order discontinuity. It should be pointed out that when every point of Γ has a first-order discontinuity, an isolated point M may have a second or higher-order discontinuity.

Let U be a function of x and t and

$$U^+ = \lim_{\xi \to 0} U(\hat{x}(t) + \xi, t),$$

$$U^- = \lim_{\xi \to 0} U(\hat{x}(t) - \xi, t), \tag{6.14}$$

$$[U] = U^- - U^+, \tag{6.15}$$

where ξ is a positive quantity. Thus U^+ (or U^-) is the value of U in front of (or behind) the elastic-plastic boundary Γ, and $[U]$ is the discontinuity of U across Γ. Applying Equations (6.8) and (6.9) to the front and back of Γ and subtracting the resulting equations, we obtain

$$[\sigma_x] = \rho[v_t], \tag{6.16}$$

$$[v_x] = \frac{1}{\rho}\left[\frac{\sigma_t}{c^2}\right]. \tag{6.17}$$

The total derivative of σ and v along Γ is

$$\left.\frac{d\sigma}{dt}\right|_\Gamma = \lambda\sigma_x + \sigma_t, \qquad \left.\frac{dv}{dt}\right|_\Gamma = \lambda v_x + v_t, \tag{6.18}$$

where σ_x, σ_t, v_x and v_t are to be evaluated on Γ. Continuity of σ and v across Γ implies that $d\sigma/dt$ and dv/dt are also continuous and hence

$$\lambda[\sigma_x] + [\sigma_t] = 0, \tag{6.19}$$

$$\lambda[v_x] + [v_t] = 0. \tag{6.20}$$

Combining Equations (6.16), (6.19) and (6.20) we obtain

$$[\sigma_t] = -\lambda[\sigma_x] = -\rho\lambda[v_t] = \rho\lambda^2[v_x]. \tag{6.21}$$

We see that if σ_t is continuous (or discontinuous), then σ_x, v_t and v_x are also continuous (or discontinuous).

Theorem 6.3.1A. *At any point M of Γ, the first-order derivatives are either all continuous or all discontinuous.*

From Equations (6.17) and (6.21) we have

$$[\sigma_t] = \lambda^2\left[\frac{\sigma_t}{c^2}\right]. \tag{6.22}$$

Equation (6.22) can be written in several different forms. We have

$$\lambda^2 = \frac{[\sigma_t]}{[\sigma_t/c^2]}, \qquad \left[\left(1 - \frac{\lambda^2}{c^2}\right)\sigma_t\right] = 0. \tag{6.23}$$

When Γ is a loading boundary, the + side is elastic while the − side is plastic. The situation is reversed when Γ is an unloading boundary. If we use the superscripts e and p for σ_t to denote the elastic and plastic regions, respectively, the second of Equations (6.23) can be written in full as

$$\left(1 - \frac{\lambda^2}{c_e^2}\right)\sigma_t^e = \left(1 - \frac{\lambda^2}{c_p^2}\right)\sigma_t^p, \tag{6.24}$$

which is equivalent to Equation (6.1).

If Γ is an unloading boundary, $\sigma_t^e \leq 0$, $\sigma_t^p \geq 0$ and the left-hand side of Equation (6.1) is negative. For the right-hand side to be negative we have

$$c_p \leq \lambda \leq c_e, \quad \text{for an unloading boundary.} \tag{6.25}$$

If Γ is a loading boundary, $\sigma_t^e \geq 0$, $\sigma_t^p \geq 0$ and the left-hand side of Equation (6.1) is positive. For the right-hand side to be positive, we have

$$\lambda \geq c_e \quad \text{or} \quad \lambda \leq c_p, \quad \text{for a loading boundary.} \tag{6.26}$$

The inequalities in Equations (6.25) and (6.26) limit the possible speed of Γ depending on whether Γ is a loading or unloading boundary.

When the point M is on an unloading boundary, $\sigma_t^e \leq 0$ and $\sigma_t^p \geq 0$ imply that $\sigma_t^e \neq \sigma_t^p$ unless $\sigma_t^e = \sigma_t^p = 0$. This means that σ_t can be continuous only when σ_t^e and σ_t^p both vanish. If M is on a loading boundary, $\sigma_t^e \geq 0$ and $\sigma_t^p \geq 0$ imply that σ_t can be continuous and nonzero. However, $\sigma_t^e = \sigma_t^p \neq 0$ means that the left-hand side of Equation (6.1) is unity. For the right-hand side to be unity we must have $c_e = c_p$ or $\lambda = 0$. Since we have assumed $\lambda \neq 0$, we have the following theorem.

Theorem 6.3.1B. *At a point M on an unloading boundary, the first-order derivatives of σ and v can be continuous only when $\sigma_t^e = \sigma_t^p = 0$. If M is on a loading boundary, the first-order derivatives of σ and v can be continuous only when $\sigma_t^e = \sigma_t^p = 0$ or when $c_e = c_p$ at M.*

We now consider the situation in which σ_t is continuous at M. By Equation (6.21) σ_x, v_t, v_x are also continuous. From Equation (6.24), $\sigma_t^e = \sigma_t^p$ leads to

$$\sigma_t^e = \sigma_t^p = 0, \tag{6.27}$$

provided $c_e \neq c_p$. By Equation (6.9), we obtain

$$v_x^e = v_x^p = 0. \tag{6.28}$$

Theorem 6.3.2. *At a point M on the elastic-plastic boundary Γ, if anyone of σ_t, σ_x, v_t, v_x is continuous, they are all continuous. Moreover, σ_t^e, σ_t^p, v_x^e and v_x^p vanish provided $c_e \neq c_p$.*

We see that Theorems 6.3.1 and 6.3.2 hold even if $\lambda = c_e$ or c_p at M.

Next, by Equation (6.9), $\sigma_t^e = 0$ implies $v_x^e = 0$ and vice versa. If $\lambda \neq c_p$, it follows from Equation (6.24) that $\sigma_t^p = 0$ and from Equation (6.9) $v_x^p = 0$. Hence, $[\sigma_t] = 0$ and, in view of Equation (6.21), we have the following theorem:

Theorem 6.3.3A. *Let $\lambda \neq c_p$ at M. If σ_t^e or v_x^e vanishes at M, σ_t^e, σ_t^p, v_x^e, v_x^p all vanish and σ_x, v_t are continuous at M.*

This theorem is useful when Γ is a loading boundary in which $\sigma_t^e(x, t)$, $v_x^e(x, t)$ are known. The vanishing of σ_t^e on the side of Γ where the solution is known assures the vanishing of σ_t^p on the other side where the solution is to be found. Similarly, the following theorem is useful when Γ is an unloading boundary.

Theorem 6.3.3B. *Let $\lambda \neq c_e$ at M. If σ_t^p or v_x^p vanishes at M, σ_t^e, σ_t^p, v_x^e, v_x^p all vanish and σ_x, v_t are continuous at M.*

We see that for $\lambda \neq c_e$ or c_p, the vanishing of σ_t^e implies the vanishing of σ_t^p and vice versa. If we wish to include the possibility of $\lambda = c_e$ or c_p, we have:

Theorem 6.3.4. *At the point M, $\sigma_t^e = \sigma_t^p = 0$ imply $v_x^e = v_x^p = 0$ and vice versa. Moreover, σ_x and v_t are continuous at M.*

In Theorems 6.3.1 to 6.3.3, we took pains to separate the restrictions $c_e \neq c_p$, $\lambda \neq c_e$ and $\lambda \neq c_p$ to obtain rigorous results. If we impose all three restrictions, these theorems can be combined and stated as follows.

Theorem 6.3.5. *At the point M where $c_e \neq c_p$, $\lambda \neq c_e$ and $\lambda \neq c_p$, the eight quantities $[\sigma_t]$, $[\sigma_x]$, $[v_t]$, $[v_x]$, σ_t^e, σ_t^p, v_x^e, v_x^p either all vanish or all do not vanish. In other words, vanishing (or nonvanishing) of one quantity implies the vanishing (or nonvanishing) of the remaining seven quantities.*

It should be pointed out that the theorems presented in this section apply to any point M on Γ. Thus, for instance, according to Theorem 6.3.5, the eight quantities $[\sigma_t]$, $[\sigma_x]$, $[v_t]$, $[v_x]$, σ_t^e, σ_t^p, v_x^e, v_x^p may all vanish simultaneously at the isolated point M on Γ and all be nonzero at neighboring points of M on Γ.

In the remainder of this paper we assume that $c_e \neq c_p$.

6.4 Discontinuities of second order at M, first order on Γ

In this section we assume that at the point M on the elastic-plastic boundary Γ, the first order derivatives of σ and v are continuous and $c_e \neq c_p$. This means, by Theorem 6.3.2

$$\sigma_t^e = \sigma_t^p = v_x^e = v_x^p = 0, \quad \text{at} \quad M. \tag{6.29}$$

Other points on Γ have discontinuous first-order derivatives. The case in which every point on Γ has a continuous first-order derivative will be discussed in the next section.

Differentiation of Equations (6.8) and (6.9) with respect to x and t yields, noticing that c is a function of σ and using Equation (6.29)

$$\begin{aligned} \sigma_{xx} &= \rho v_{xt} \\ \sigma_{xt} &= \rho v_{tt} \\ v_{xx} &= \frac{\sigma_{xt}}{\rho c^2} \end{aligned} \tag{6.30}$$

$$v_{xt} = \frac{\sigma_{tt}}{\rho c^2}.$$

This can be written in terms of σ_{tt} and v_{tt} as

$$\begin{aligned}
\sigma_{xx} &= \frac{\sigma_{tt}}{c^2} \\
\sigma_{xt} &= \rho v_{tt} \\
v_{xx} &= \frac{v_{tt}}{c^2} \\
v_{xt} &= \frac{\sigma_{tt}}{\rho c^2}.
\end{aligned} \tag{6.31}$$

The second order total derivatives of σ and v along the boundary Γ are

$$\begin{aligned}
\left.\frac{\mathrm{d}^2\sigma}{\mathrm{d}t^2}\right|_\Gamma &= \left(\lambda\frac{\partial}{\partial x} + \frac{\partial}{\partial t}\right)^2 \sigma \\
&= \lambda^2\sigma_{xx} + 2\lambda\sigma_{xt} + \sigma_{tt} + \lambda_t\sigma_x, \\
\left.\frac{\mathrm{d}^2 v}{\mathrm{d}t^2}\right|_\Gamma &= \left(\lambda\frac{\partial}{\partial x} + \frac{\partial}{\partial t}\right)^2 v \\
&= \lambda^2 v_{xx} + 2\lambda v_{xt} + v_{tt} + \lambda_t v_x.
\end{aligned} \tag{6.32}$$

The continuity of σ and v along Γ implies that $\mathrm{d}^2\sigma/\mathrm{d}t^2$ and $\mathrm{d}^2v/\mathrm{d}t^2$ are also continuous. Hence,

$$\begin{aligned}
\lambda^2[\sigma_{xx}] + 2\lambda[\sigma_{xt}] + [\sigma_{tt}] &= 0, \\
\lambda^2[v_{xx}] + 2\lambda[v_{xt}] + [v_{tt}] &= 0,
\end{aligned} \tag{6.33}$$

or, using Equation (6.31)

$$\left[\left(1 + \frac{\lambda^2}{c^2}\right)\sigma_{tt}\right] + 2\rho\lambda[v_{tt}] = 0, \tag{6.34a}$$

$$\left[\left(1 + \frac{\lambda^2}{c^2}\right)v_{tt}\right] + \frac{2\lambda}{\rho}\left[\frac{\sigma_{tt}}{c^2}\right] = 0. \tag{6.34b}$$

Suppose that, at the point M, σ_{tt} and v_{tt} are continuous. Since it is assumed that $c_e \neq c_p$, Equation (6.34a) reduces to

$$\sigma_{tt}^e = \sigma_{tt}^p = 0, \tag{6.35}$$

and Equation (6.34b) yields

$$v_{tt}^e = v_{tt}^p = 0. \tag{6.36}$$

Use of Equation (6.31) leads to the following theorem.

Theorem 6.4.1. *Let σ_{tt} and v_{tt} be continuous at M. Then all second-order derivatives of σ and v vanish at M.*

We conclude that if M has third-order discontinuities, all second-order derivatives are continuous by definition and hence are zero by Theorem 6.4.1.

We could assume that Equation (6.35) holds at M. Equation (6.34a) shows that v_{tt} is continuous and thus both σ_{tt} and v_{tt} are continuous. In view of Theorem 6.4.1, we have:

Theorem 6.4.2. *Let $\sigma_{tt}^e = \sigma_{tt}^p = 0$ at M. Then all second-order derivatives of σ and v vanish at M.*

It should be pointed out that Theorems 6.4.1 and 6.4.2 hold without the assumption that $\lambda \neq c_e, c_p$. Moreover, the assumptions in the theorems are imposed only at the isolated point M.

Instead of assuming $\sigma_{tt}^e = \sigma_{tt}^p = 0$ at M, we could assume that $\sigma_{tt}^e = v_{tt}^e = 0$ at M. Equations (6.34), after eliminating v_{tt}^p between them then yield

$$\left(1 - \frac{\lambda^2}{c_p^2}\right)\sigma_{tt}^p = 0. \tag{6.37}$$

Therefore, if $\lambda \neq c_p$, $\sigma_{tt}^p = 0$ and by Equation (6.34a), $v_{tt}^p = 0$. Thus, σ_{tt} and v_{tt} are continuous and by Theorem 6.4.1, all second-order derivatives vanish. Similar results are obtained if $\sigma_{tt}^p = v_{tt}^p = 0$ and $\lambda \neq c_e$.

Theorem 6.4.3. *At the point M, let $\sigma_{tt}^e = v_{tt}^e = 0$, $\lambda \neq c_p$, or $\sigma_{tt}^p = v_{tt}^p = 0$, $\lambda \neq c_e$. Then all second-order derivatives of σ and v vanish at M.*

Using Equation (6.31), ρv_{tt} may be replaced by σ_{xt}. Equation (6.34a) can then be written as

$$\left[\left(1 + \frac{\lambda^2}{c^2}\right)\sigma_{tt}\right] + 2\lambda[\sigma_{xt}] = 0. \tag{6.38}$$

This is Equation (6.3b). By a linear combination of Equations (6.34a) and (6.34b) and with ρv_{tt} replaced by σ_{xt}, we have

$$\left[\left(1 - \frac{\lambda^2}{c^2}\right)(\sigma_{tt} + \lambda\sigma_{xt})\right] = 0. \tag{6.39}$$

Written in full, this takes the form of the following expression:

$$\frac{\sigma_{tt}^{p} + \lambda\sigma_{xt}^{p}}{\sigma_{tt}^{e} + \lambda\sigma_{xt}^{e}} = \frac{\dfrac{1}{c_e^2} - \dfrac{1}{\lambda^2}}{\dfrac{1}{c_p^2} - \dfrac{1}{\lambda^2}}, \tag{6.40}$$

which is Equation (6.3a). Since the total derivative of σ_t along Γ is

$$\left(\frac{\mathrm{d}}{\mathrm{d}t}\sigma_t\right)\bigg|_{\Gamma} = \sigma_{tt} + \lambda\sigma_{xt}, \tag{6.41}$$

we may write Equations (6.3.8) and (6.39) as

$$2\left[\left(\frac{\mathrm{d}}{\mathrm{d}t}\sigma_t\right)\bigg|_{\Gamma}\right] - \left[\left(1 - \frac{\lambda^2}{c^2}\right)\sigma_{tt}\right] = 0, \quad \left[\left(1 - \frac{\lambda^2}{c^2}\right)\left(\frac{\mathrm{d}}{\mathrm{d}t}\sigma_t\right)\bigg|_{\Gamma}\right] = 0, \tag{6.42}$$

and Equation (6.40) as

$$\frac{\left(\dfrac{\mathrm{d}}{\mathrm{d}t}\sigma_t^{p}\right)\bigg|_{\Gamma}}{\left(\dfrac{\mathrm{d}}{\mathrm{d}t}\sigma_t^{e}\right)\bigg|_{\Gamma}} = \frac{\dfrac{1}{c_e^2} - \dfrac{1}{\lambda^2}}{\dfrac{1}{c_p^2} - \dfrac{1}{\lambda^2}}. \tag{6.43}$$

At an isolated point M which has second-order discontinuities and $c_e \neq c_p$, σ_{tt} and v_{tt} at M are related by Equations (6.34), (6.38), (6.39) or (6.42) while other second-order derivatives are related to σ_{tt} and v_{tt} through Equation (6.31). Examples in which Γ has first-order discontinuities except at the isolated point M, where it has second-order discontinuities, can be found in [22].

6.5 Discontinuities of second order at M and on Γ

We now consider the situation in which *the first order derivatives of σ and v are continuous at every point of the elastic-plastic boundary Γ which contains the point M and $c_e \neq c_p$.* Equations (6.29) then hold on Γ and we have

$$\begin{aligned} 0 &= \left(\frac{\mathrm{d}}{\mathrm{d}t}\sigma_t\right)\bigg|_{\Gamma} = \sigma_{tt} + \lambda\sigma_{xt}, \\ 0 &= \left(\frac{\mathrm{d}}{\mathrm{d}t}v_x\right)\bigg|_{\Gamma} = v_{xt} + \lambda v_{xx}. \end{aligned} \tag{6.44}$$

The second of Equation (6.42) is automatically satisfied and the first of Equation (6.42) reduces to

$$\left[\left(1 - \frac{\lambda^2}{c^2}\right)\sigma_{tt}\right] = 0. \tag{6.45}$$

Written in full, we have

$$\left(1 - \frac{\lambda^2}{c_e^2}\right)\sigma_{tt}^e = \left(1 - \frac{\lambda^2}{c_p^2}\right)\sigma_{tt}^p, \quad \text{or} \quad \frac{\sigma_{tt}^p}{\sigma_{tt}^e} = \frac{\dfrac{1}{c_e^2} - \dfrac{1}{\lambda^2}}{\dfrac{1}{c_p^2} - \dfrac{1}{\lambda^2}}, \tag{6.46}$$

which is Equation (6.2). Using a similar argument in deriving Equations (6.25) and (6.26), we obtain the following results. If Γ is an unloading boundary, $\sigma_{tt}^e \leq 0$, $\sigma_{tt}^p \leq 0$ and we have

$$\lambda \geq c_e \quad \text{or} \quad \lambda \leq c_p, \quad \text{for an unloading boundary.} \tag{6.47}$$

If Γ is a loading boundary, $\sigma_{tt}^e \leq 0$, $\sigma_{tt}^p \geq 0$ and we obtain

$$c_p \leq \lambda \leq c_e \quad \text{for a loading boundary.} \tag{6.48}$$

The inequalities in Equations (6.47) and (6.48) are identical to those in Equations (6.25) and (6.26) except that the conditions applied to the loading and unloading boundary are interchanged.

If we write Equation (6.45) as

$$[\sigma_{tt}] = \lambda^2\left[\frac{\sigma_{tt}}{c^2}\right],$$

use of Equations (6.30) and (6.44) leads to

$$-\rho\lambda[v_{tt}] = -\lambda[\sigma_{xt}] = [\sigma_{tt}] = \lambda^2\left[\frac{\sigma_{tt}}{c^2}\right] = \rho\lambda^2[v_{xt}]$$

$$= -\rho\lambda^3[v_{xx}] = \lambda^2[\sigma_{xx}]. \tag{6.49}$$

Therefore, the second-order derivatives of σ and v are either all continuous or all discontinuous. In view of the similarity between Equations (6.46) and (6.24), we have the following theorem [21] similar to Theorems 6.3.1A and 6.3.1B except that we have assumed $c_e \neq c_p$ in this section.

Theorem 6.5.1. *At any point of Γ, the second-order derivatives of σ and v are either all continuous or all discontinuous. Moreover, they can be continuous only when $\sigma_{tt}^e = \sigma_{tt}^p = 0$.*

Equations (6.31) indicate that all second-order derivatives can be expressed in terms of σ_{tt} and v_{tt}. Substituting into the first equation of (6.44) we obtain

$$\sigma_{tt} = -\rho\lambda v_{tt}. \tag{6.50}$$

Substitution in the second equation of (6.44) leads to the same equation. Therefore, all second-order derivatives can be expressed in terms of σ_{tt}. In view of Equation (6.50), Theorem 6.4.1 can be modified for the present case as follows:

Theorem 6.5.2. *Let σ_{tt} be continuous at M. Then all second-order derivatives of σ and v vanish at M.*

Notice that Theorem 6.5.2 does not require that $\lambda \neq c_e$ or c_p, nor $\sigma_{tt}^e = \sigma_{tt}^p = 0$. All it requires is that σ_{tt} be continuous. However, if $\sigma_{tt}^e = 0$ and $\lambda \neq c_p$ at M, Equation (6.46) yields $\sigma_{tt}^p = 0$ and hance σ_{tt} is continuous. Likewise, if $\sigma_{tt}^p = 0$ and $\lambda \neq c_e$ at M, $\sigma_{tt}^e = 0$ and hence σ_{tt} is continuous at M.

Theorem 6.5.3. *Let $\sigma_{tt}^e = 0$, $\lambda \neq c_p$, or $\sigma_{tt}^p = 0$, $\lambda \neq c_e$ at M. Then all second-order derivatives of σ and v vanish at M.*

We see that if $\lambda \neq c_e$ or c_p, $\sigma_{tt}^e = 0$ implies $\sigma_{tt}^p = 0$ and vice versa. This conclusion can also be reached by observation of Equation (6.46).

6.6 Discontinuities of third and higher order at M, first order on Γ

For the third and higher order discontinuities, it is more convenient to use the matrix expressions in Equations (6.10) and (6.11) instead of Equations (6.8) and (6.9). We will first re-derive some of the results in Section 6.4 in matrix notation. This means that we assume w_x and w_t are continuous and Equations (6.29) hold at M. Differentiation of Equation (6.10) with respect to t and x yields

$$\begin{aligned} w_{xt} &= Nw_{tt} + N_t w_t \\ w_{xx} &= Nw_{xt} + N_x w_t. \end{aligned} \tag{6.51}$$

With N and w defined in Equation (6.11) and $\sigma_t^e = \sigma_t^p = 0$ by Equation (6.29),

$$N_t w_t = \begin{bmatrix} 0 & \left(\dfrac{1}{\rho c^2}\right)_t \\ 0 & 0 \end{bmatrix} \begin{bmatrix} v_t \\ 0 \end{bmatrix} = 0 \quad \text{at} \quad M. \tag{6.52}$$

Similarly,

$$N_x w_t = 0 \quad \text{at} \quad M. \tag{6.53}$$

Equations (6.51) reduce to

$$\begin{aligned} w_{xt} &= Nw_{tt} \\ w_{xx} &= N^2 w_{tt} \end{aligned} \tag{6.54}$$

at M. This is Equation (6.31). The second-order total derivative of w along Γ is

$$\left.\frac{d^2w}{dt^2}\right|_\Gamma = \left(\lambda\frac{\partial}{\partial x} + \frac{\partial}{\partial t}\right)^2 w$$

$$= \lambda^2 w_{xx} + 2\lambda w_{xt} + w_{tt} + \lambda_t w_x. \tag{6.55}$$

Since w_x and d^2w/dt^2 are continuous at M we have

$$\lambda^2[w_{xx}] + 2\lambda[w_{xt}] + [w_{tt}] = 0, \tag{6.56}$$

or using Equation (6.54),

$$[(\lambda N + I)^2 w_{tt}] = 0, \tag{6.57}$$

where I is the identity matrix. Equations (6.56) and (6.57) are identical to Equations (6.33) and (6.34), respectively.

We now consider the case in which Γ *has first-order discontinuities while at an isolated point M it has nth order discontinuities.* This means, by definition, that all derivatives of w of order up to $(n - 1)$ are continuous at M. For $n = 3$ this also means that by Theorems 6.3.2 and 6.4.1 all first- and second-order derivatives of w vanish at M with the exception of σ_x and v_t. We will assume that *with the exception of* σ_x *and* v_t *all derivatives of w of order up to* $(n - 1)$ *vanish at M.* The assumption is true for $n = 3$. We will see later (Theorem 6.6.1B) that the assumption is true for any $n \geq 3$. Incidentally, the following derivations though intended for $n \geq 3$ are also valid for $n = 2$.

Following the procedure in deriving Equations (6.54), we have

$$\begin{aligned} \frac{\partial^n w}{\partial x \partial t^{n-1}} &= N\frac{\partial^n w}{\partial t^n}, \\ &\cdots\cdots \\ \frac{\partial^n w}{\partial x^k \partial t^{n-k}} &= N^k\frac{\partial^n w}{\partial t^n}, \\ &\cdots\cdots \\ \frac{\partial^n w}{\partial x^n} &= N^n\frac{\partial^n w}{\partial t^n}. \end{aligned} \tag{6.58}$$

Hence, all nth order derivatives can be expressed in terms of $\partial^n w/\partial t^n$. The nth total derivative of w along Γ is, using Equation (6.58),

$$\left.\frac{d^n w}{dt^n}\right|_\Gamma = \left(\lambda\frac{\partial}{\partial x} + \frac{\partial}{\partial t}\right)^n w$$

$$= \sum_{k=1}^{n} {}_nC_k \lambda^k N^k \frac{\partial^n w}{\partial t^n} + \frac{d^{n-1}\lambda}{dt^{n-1}} w_x$$

$$= (\lambda N + I)^n \frac{\partial^n w}{\partial t^n} + \frac{d^{n-1}\lambda}{dt^{n-1}} w_x$$

$$= Q^n \frac{\partial^n w}{\partial t^n} + \frac{d^{n-1}\lambda}{d t^{n-1}} w_x, \tag{6.59}$$

in which the binomial coefficients ${}_nC_k$ and Q are given by

$${}_nC_k = \frac{n!}{k!(n-k)!}, \tag{6.60}$$

$$Q = (\lambda N + I). \tag{6.61}$$

The continuity of $d^n w/dt^n$ across Γ, noticing that w_x is continuous at M, leads to

$$\left[Q^n \frac{\partial^n w}{\partial t^n} \right] = 0. \tag{6.62}$$

This generalizes the result for $n = 2$ in Equation (6.57).

The fact that N is a 2×2 matrix implies that Q^n can be expressed as a linear combination of I and N. Indeed, carrying out the binomial expansion of $(\lambda N + I)^n$ and observing that

$$N^2 = \frac{1}{c^2} I, \tag{6.63}$$

we have

$$Q^n = \phi_n I + \psi_n c N = \begin{bmatrix} \phi_n & \dfrac{\gamma \psi_n}{\rho \lambda} \\ \dfrac{\rho \lambda \psi_n}{\gamma} & \phi_n \end{bmatrix}, \tag{6.64}$$

where [19],

$$\begin{aligned} \phi_n &= \sum_{k=0,2,\dots} {}_nC_k \gamma^k = \frac{1}{2}\{(1+\gamma)^n + (1-\gamma)^n\}, \\ \psi_n &= \sum_{k=1,3,\dots} {}_nC_k \gamma^k = \frac{1}{2}\{(1+\gamma)^n - (1-\gamma)^n\}, \end{aligned} \tag{6.65}$$

$$\gamma = \frac{\lambda}{c}. \tag{6.66}$$

From the determinant

$$\|Q\| = (1-\gamma^2), \quad \|Q^n\| = (1-\gamma^2)^n, \tag{6.67}$$

we see that Q^n is nonsingular unless $\lambda = c$. With Q^n given by Equations (6.64), Equation (6.62) can be written as [21],

$$\left[\phi_n \frac{\partial^n v}{\partial t^n} + \frac{\gamma \psi_n}{\rho \lambda} \frac{\partial^n \sigma}{\partial t^n} \right] = 0, \quad \left[\rho \lambda \frac{\psi_n}{\gamma} \frac{\partial^n v}{\partial t^n} + \phi_n \frac{\partial^n \sigma}{\partial t^n} \right] = 0. \tag{6.68}$$

This is the generalization of Equations (6.34) for $n \geq 2$.

Let $\partial^n w/\partial t^n$ be continuous at M. Equation (6.62) becomes

$$[Q^n]\frac{\partial^n w}{\partial t^n} = 0. \tag{6.69}$$

We therefore have the theorem following Equations (6.69) and (6.58):

Theorem 6.6.1A. *Let $\partial^n w/\partial t^n$ be continuous at M. Then all nth order derivatives of w vanish at M unless $[Q^n]$ is singular at M.*

For $n = 2$, it can easily be shown that $[Q^2]$ is nonsingular and hence Theorem 6.4.1 holds. We are not able to show whether $[Q^n]$ is nonsingular for any integer n and for any value of λ. However, even if $[Q^n]$ is singular for a certain n, it can occur only for a specific value (or values) of λ. Therefore, $[Q^n]$ is not likely to be singular. The following theorem does not require the assumption that $[Q^n]$ be nonsingular.

Theorem 6.6.1B. *Let $\partial^n w/\partial t^n$ and $\partial^n w/\partial x^2\partial t^{n-2}$ be continuous at M. Then all nth order derivatives of w vanish at M.*

The proof is immediate if we use the following identity which is taken from Equations (6.58) making use of Equation (6.63):

$$\frac{\partial^n w}{\partial x^2\partial t^{n-2}} = \frac{1}{c^2}\frac{\partial^n w}{\partial t^n}.$$

The continuity assumption of the theorem implies that

$$0 = \left[\frac{1}{c^2}\right]\frac{\partial^n w}{\partial t^n}.$$

Since we have assumed $c_e \neq c_p$, $\partial^n w/\partial t^n$ vanishes and the theorem follows in view of Equations (6.58).

From Theorem 6.6.1B we verify the assumption made earlier that continuity of all nth-order derivatives of w at M implies the vanishing of all nth order derivatives of w at M.

Instead of assuming the continuity of $\partial^n w/\partial t^n$ in Theorem 6.6.1A, we may assume that $\partial^n w^e/\partial t^n = 0$ at M. Equation (6.62) then reduces to

$$(Q^n)^p\frac{\partial^n w^p}{\partial t^n} = 0. \tag{6.70}$$

If $\lambda \neq c_p$, $(Q^n)^p$ is nonsingular and $\partial^n w^p/\partial t^n = 0$. It follows from Equations (6.58) that all nth-order derivatives of w vanish. Similar results are obtained if $\partial^n w^p/\partial t^n = 0$ and $\lambda \neq c_e$ at M.

Theorem 6.6.2. *At the point M, let $\partial^n w^e/\partial t^n = 0$, $\lambda \neq c_p$, or $\partial^n w^p/\partial t^n = 0$, $\lambda \neq c_e$. Then all nth-order derivatives of w vanish.*

Unlike Theorem 6.6.1A, we do not require that $[Q^n]$ be nonsingular.

Finally, consider the case $\partial^n \sigma^e/\partial t^n = \partial^n \sigma^p/\partial t^n = 0$. Equation (6.62) with the use of Equation (6.64) becomes

$$\begin{bmatrix} \phi_n^e & -\phi_n^p \\ \dfrac{\psi_n^e}{\gamma^e} & -\dfrac{\psi_n^p}{\gamma^p} \end{bmatrix} \begin{bmatrix} \dfrac{\partial^n v^e}{\partial t^n} \\ \dfrac{\partial^n v^p}{\partial t^n} \end{bmatrix} = 0. \tag{6.71}$$

The coefficient matrix is nonsingular if

$$\left(\frac{\gamma\phi_n}{\psi_n}\right)^e \neq \left(\frac{\gamma\phi_n}{\psi_n}\right)^p. \tag{6.72}$$

It can be shown that $\gamma\phi_n/\psi_n$ and its first-order derivative with respect to γ are positive for positive γ. Therefore, $\gamma\phi_n/\psi_n$ is a monotonically increasing function of γ and Equation (6.72) holds provided $c_e \neq c_p$ which we have assumed.

Theorem 6.6.3. *Let $\partial^n\sigma^e/\partial t^n = \partial^n\sigma^p/\partial t^n = 0$ at M. Then all nth-order derivatives of w vanish at M.*

Theorems 6.6.2 and 6.6.3 generalize Theorems 6.4.3 and 6.4.2, respectively, for $n > 2$.

6.7 Discontinuities of third and higher order at M, second order on Γ

In this section we assume that, *at every point on Γ, the first-order derivatives of w are continuous except at an isolated point M where the first $(n-1)$ derivatives are continuous.* Again, our objective is for $n \geq 3$ but the results obtained below also apply to $n = 2$. We then have, in addition to the results obtained in Section 6.6, $\sigma_t = 0$ and $v_x = 0$ on Γ according to Equation (6.29). $\sigma_t = 0$ can now be written as

$$Jw_t = 0, \quad J = \begin{bmatrix} 0 & 1 \\ 0 & 0 \end{bmatrix}. \tag{6.73}$$

The fact that $\sigma_t = 0$ on Γ implies that

$$0 = \frac{d^{n-1}}{dt^{n-1}}(Jw_t)\bigg|_\Gamma = J\left(\lambda\frac{\partial}{\partial x} + \frac{\partial}{\partial t}\right)^{n-1} w_t$$

$$= JQ^{n-1}\frac{\partial^n w}{\partial t^n}, \tag{6.74}$$

where use has been made of Equations (6.58) and (6.61). Carrying out the matrix multiplication, we have [21]

$$\rho\lambda\psi_{n-1}\frac{\partial^n v}{\partial t^n} + \gamma\phi_{n-1}\frac{\partial^n \sigma}{\partial t^n} = 0. \tag{6.75}$$

This reduces to Equation (6.50) for $n = 2$. As to $v_x = 0$ on Γ, we may write it as

$$J^T w_x = 0, \quad J^T = \begin{bmatrix} 0 & 0 \\ 1 & 0 \end{bmatrix}. \tag{6.76}$$

One then has

$$0 = \frac{d^{n-1}}{dt^{n-1}}(J^T w_x)\bigg|_\Gamma = J^T\left(\lambda\frac{\partial}{\partial x} + \frac{\partial}{\partial t}\right)^{n-1} w_x$$

$$= J^T Q^{n-1} N \frac{\partial^n w}{\partial t^n}. \tag{6.77}$$

It is readily shown that Equation (6.77) leads to (6.75) and hence no new result is obtained by considering $v_x = 0$ on Γ.

There are two equations in Equation (6.62) which can be reduced to one by using Equation (6.75) or its original forms in Equations (6.74) or (6.77). Using the relation

$$\lambda N = Q - I,$$

which is obtained from Equation (6.61), Equation (6.77) can be rewritten as

$$J^T Q^n \frac{\partial^n w}{\partial t^n} = J^T Q^{n-1} \frac{\partial^n w}{\partial t^n}. \tag{6.78}$$

Equation (6.62) then becomes

$$\left[J^T Q^{n-1} \frac{\partial^n w}{\partial t^n}\right] = 0, \tag{6.79}$$

which is, upon carrying out the matrix multiplication [19]

$$\left[\rho\lambda\phi_{n-1}\frac{\partial^n v}{\partial t^n} + \gamma\psi_{n-1}\frac{\partial^n \sigma}{\partial t^n}\right] = 0. \tag{6.80}$$

Substituting $\partial^n v/\partial t^n$ from Equation (6.75), we finally have

$$\left[q_n \frac{\partial^n \sigma}{\partial t^n}\right] = 0, \quad \text{or} \quad \frac{\dfrac{\partial^n \sigma^p}{\partial t^n}}{\dfrac{\partial^n \sigma^e}{\partial t^n}} = \frac{q_n^e}{q_n^p}, \tag{6.81}$$

where

$$q_n = \frac{\gamma(\phi_{n-1}^2 - \psi_{n-1}^2)}{\psi_{n-1}}. \tag{6.82}$$

For $n = 2$, Equation (6.81) reduces to Equation (6.45) and the second of Equation (6.46), respectively.

Thus, when Γ has the continuous first order derivatives of w, the two equations in Equations (6.62) or (6.68) are replaced by Equation (6.75) and either Equations (6.80), or (6.81). Theorems 6.6.1 to 6.6.3 in Section 6.6 can be modified with fewer restrictions. For Theorem 6.6.1A, we obtain from Equations (6.81), (6.75) and (6.58) the following theorem.

Theorem 6.7.1A. *Let $\partial^n\sigma/\partial t^n$ be continuous at M. Then all nth-order derivatives of w vanish at M unless $[q_n]$ vanishes at M.*

As in the discussion of $[Q^n]$ after Theorem 6.6.1A, it is unlikely that $[q_n]$ vanishes at M. A counterpart to Theorem 6.6.1B is the next theorem which does not require that $[q_n]$ be nonzero.

Theorem 6.7.1B. *At the point M, if $\partial^n w/\partial t^n$ is continuous, all nth-order derivatives of w vanish at M.*

To prove the theorem, we write Equation (6.75) as

$$\rho\lambda\left[\frac{\partial^n v}{\partial t^n}\right] + \left[\frac{\gamma\phi_{n-1}}{\psi_{n-1}}\frac{\partial^n \sigma}{\partial t^n}\right] = 0,$$

and the continuity of $\partial^n w/\partial t^n$ means that

$$\left[\frac{\gamma\phi_{n-1}}{\psi_{n-1}}\right]\frac{\partial^n \sigma}{\partial t^n} = 0.$$

Since $\gamma\phi_n/\psi_n$ is a monotonically increasing function of γ as we stated earlier, we have $\partial^n\sigma/\partial t^n = 0$ and Theorem 6.7.1B follows from Equations (6.75) and (6.58).

Next, let $\partial^n\sigma^e/\partial t^n = 0$. Equation (6.81) yields $\partial^n\sigma^p/\partial t^n = 0$ because q_n^p is nonzero provided $\lambda \neq c_p$. Equations (6.75) and (6.58) then show that all nth-order derivatives of w vanish. Similar results are obtained if $\partial^n\sigma^p/\partial t^n = 0$ and $\lambda \neq c_e$.

Theorem 6.7.2. *At the point M, let $\partial^n\sigma^e/\partial t^n = 0$, $\lambda \neq c_p$, or $\partial^n\sigma^p/\partial t^n = 0$, $\lambda \neq c_e$. Then all nth-order derivatives of w vanish at M.*

6.8 Impossibility of third and higher order discontinuities on Γ

In this section, we consider the situation in which the first- and second-order derivatives of w are continuous on every point of Γ on which $c_e \neq c_p$. From Theorems 6.3.2 and 6.5.2, all first- and second-order derivatives of w vanish except σ_x and v_t. Using these properties, we obtain by differentiating the first of Equation (6.51) with respect to t to obtain

$$w_{xtt} = N w_{ttt}, \tag{6.83}$$

where use has been made of Equations (6.52) and (6.53). On the other hand, $w_{tt} = 0$ on Γ means

$$0 = \left.\left(\frac{\mathrm{d}}{\mathrm{d}t}w_{tt}\right)\right|_{\Gamma} = \lambda w_{xtt} + w_{ttt}. \tag{6.84}$$

Substitution of w_{xtt} from Equation (6.83) yields

$$Q w_{ttt} = 0, \tag{6.85}$$

where Q is defined in Equation (6.61). Since Q is nonsingular unless $\lambda = c$, $w_{ttt} = 0$ and by Equation (6.58) all third-order derivatives of w vanish. With all third-order derivatives of w being zero, double differentiation of the first part of Equation (6.51) with respect to t leads to

$$w_{xttt} = N w_{tttt}. \tag{6.86}$$

The vanishing of w_{ttt} on Γ implies that

$$0 = \left.\left(\frac{\mathrm{d}}{\mathrm{d}t} w_{ttt}\right)\right|_{\Gamma} = \lambda w_{xttt} + w_{tttt}. \tag{6.87}$$

We again obtain

$$Q w_{tttt} = 0. \tag{6.88}$$

Assuming $\lambda \neq c$, w_{tttt} vanishes, so do other fourth-order derivatives by Equations (6.58). The same conclusion is reached for the higher order derivatives.

Theorem 6.8.1. *A discontinuity of order higher than two cannot occur on a finite segment of* Γ *unless* $\lambda = c_e$ *or* c_p.

6.9 Improbability of second and higher order discontinuities on a loading boundary

We see that the discontinuities of second order are the highest order possible across a finite segment of an elastic-plastic boundary Γ for which $\lambda \neq c_e$ or c_p. Consider that Γ is an unloading boundary and the first order derivatives of w are continuous on Γ. Let $\sigma(x, t)$ be known in the plastic region. Since σ_t vanishes on Γ one may obtain the equation for Γ by the condition

$$\sigma_t(x, t)\Big|_{\Gamma} = 0. \tag{6.89}$$

This yields, say $t = \hat{t}(x)$ and the stress $\sigma(x, t)$ on Γ provides the new yield stress σ_Y:

$$\sigma(x, \hat{t}(x)) = \sigma_Y(x). \tag{6.90}$$

The situation is quite different if Γ is a loading boundary. While the equation for Γ is still given by Equation (6.89), the left-hand side of (6.90) must now satisfy the prescribed function $\sigma_Y(x)$. For an arbitrarily prescribed $\sigma_Y(x)$, it is not likely that Equation (6.90) is satisfied for every x on Γ.

To be more specific, we differentiate Equation (6.90) with respect to x and use Equation (6.89) to obtain

$$\sigma_x = \frac{\mathrm{d}}{\mathrm{d}x}\sigma_Y(x), \quad \text{on } \Gamma. \tag{6.91}$$

Differentiating Equation (6.90) twice, we obtain

$$\sigma_{xx} + \frac{2}{\lambda}\sigma_{xt} + \frac{1}{\lambda^2}\sigma_{tt} = \frac{\mathrm{d}^2}{\mathrm{d}x^2}\sigma_Y, \quad \text{on } \Gamma.$$

Since $\sigma_{xx} = \sigma_{tt}/c^2$ and $\sigma_{xt} = -\sigma_{tt}/\lambda$ by Equations (6.31) and (6.50), this can be written as

$$\left(\frac{1}{c^2} - \frac{1}{\lambda^2}\right)\sigma_{tt} = \frac{\mathrm{d}^2}{\mathrm{d}x^2}\sigma_Y, \quad \text{on } \Gamma. \tag{6.92}$$

If Γ is a loading boundary, regardless of whether we consider the elastic or the plastic side of Γ, we can show from Equation (6.48) that the left-hand side of Equation (6.92) is positive and nonzero. It is nonzero because $\lambda \neq c_e$ or c_p, and with $c_e \neq c_p$, $\sigma_{tt}^e = 0$ would imply that $\sigma_{tt}^p = 0$ and vice versa. By Theorem 6.5.3 there would be no discontinuities of second-order across Γ. Therefore, for a loading boundary to have a second-order discontinuity, the yield stress σ_Y must satisfy the relation

$$\frac{\mathrm{d}^2}{\mathrm{d}x^2}\sigma_Y(x) > 0 \quad \textit{on} \text{ a loading boundary.} \tag{6.93}$$

Theorem 6.9.1. *Across a loading boundary Γ for which $\lambda \neq c_e$ or c_p, a discontinuity of order higher than one cannot occur along a finite segment of Γ unless* $\mathrm{d}^2\sigma_Y/\mathrm{d}x^2 > 0$.

It should be stressed that satisfaction of Equation (6.93) does not assure the existence of second-order discontinuities on the loading boundary. Equation (6.90) still has to be satisfied for every x on Γ. An interesting point to observe is that Equation (6.92) applies to an unloading boundary also. Using Equation (6.47), one can show that on an unloading boundary,

$$\begin{aligned} \frac{\mathrm{d}^2}{\mathrm{d}x^2}\sigma_Y(x) &> 0, \text{ if } \lambda < c_p, \\ &< 0, \text{ if } \lambda > c_e. \end{aligned} \tag{6.94}$$

We therefore have the following theorem:

Theorem 6.9.2. *Let Γ be an unloading boundary with continuous first-order derivatives of w. If $\lambda > c_e$, the order of discontinuity at the subsequent loading boundary cannot be larger than one.*

6.10 Concluding remarks

We have analyzed systematically the discontinuities in the derivatives of stress and velocity across an elastic-plastic boundary for one-dimensional wave propagation in a rod. A highlight of the results is the one discovered by Yu, Wang and Zhu that discontinuities of order higher than two cannot occur on a finite segment of an elastic-plastic boundary unless the boundary coincides with the characteristics of the elastic or plastic region. Implicit in the assumption is that $c_e \neq c_p$ along the boundary. One could construct a solution which has a discontinuity in the third-order derivatives along an elastic-plastic boundary which is not a characteristic of the elastic or the plastic region but on which $\sigma = \sigma_0$ and $c_e = c_p$. For this to be possible, the stress-strain law Equation (6.4b) must be such that $\mathrm{d}f/\mathrm{d}\varepsilon = E$, $\mathrm{d}^2 f/\mathrm{d}\varepsilon^2 = 0$ and $\mathrm{d}^3 f/\mathrm{d}\varepsilon^3 < 0$ at $\varepsilon = \varepsilon_0$. It is hoped that the analysis presented here can be extended to elastic-plastic waves of combined stress [23 – 25] and to three-dimensional elastic-plastic waves [26 – 29].

References

[1] VON KARMAN, Th., 'On the Propagation of Plastic Deformation in Solids', NDRC Report, A-29 (OSRD No. 365), 1942.

[2] VON KARMAN, Th., and DUWEZ, P., 'The Propagation of Plastic Deformation in Solids', *J. Appl. Phys.*, Vol. 21, 1950, pp. 987–994.

[3] TAYLOR, G.I., 'The Plastic Wave in a Wire Extended by an Impact Load', Britsh Official Report, R.C. 329, 1942.

[4] TAYLOR, G.I., 'The Plastic Wave in a Wire Extended by an Impact Load', *The Scientific Papers of G.I. Taylor, Vol. I, Mechanics of Solids,* G.K. Batchelor, ed., University Press, Cambridge, 1958, pp. 467–479.

[5] RAKHMATULIN, Kh.A., 'The Propagation of an Unloading Wave', *Prik. Mat. Mekh.,* Vol. 9, 1945, pp. 91–100 (in Russian).

[6] RAKHMATULIN, Kh.A., 'High Speed Oblique Impact on a Flexible String Under Tension', *Prik. Mat. Mekh.,* Vol. 9, 1945, pp. 449–462 (in Russian).

[7] DONNELL, L.H., 'Longitudinal Wave Transmission and Impact', *Trans. ASME,* Vol. 52, 1930, pp. 153–161.

[8] CRISTESCU, N., *Dynamic Plasticity,* North-Holland Publ. Co., Amsterdam, 1967.

[9] BOHNENBLUST, H.F., CHARYK, J.V. and HYERS, D.H., 'Graphical Solution for Problems of Strain Propagation in Tension', NDRC Report No. A-131 (OSRD No. 1204), 1943.

[10] SHAPIRO, G.S., 'Longitudinal Vibrations of Bars', *Prik. Mat. Mekh.,* Vol. 10, 1946, pp. 597–616.

[11] WHITE, M.P. and GRIFFIS, L. VAN, 'The Permanent Strain in a Uniform Bar Due to Longitudinal Impact', *J. Appl. Mech.,* Vol. 14, 1947, pp. 337–343.

[12] RAKHMATULIN, Kh.A. and SHAPIRO, G.S., 'On the Propagation of Plane Elastic-Plastic Waves', *Prik. Mat. Mekh.,* Vol. 12, 1948, pp. 369–374.

[13] LEE, E.H., 'A Boundary Value Problem in the Theory of Plastic Wave Propagation', *Q. Appl. Math.,* Vol. 10, 1953, pp. 335–346.

[14] SALVADORI, M.G., SKALAK, R. and WEIDLINGER, P., 'Waves and Shocks in Locking and Dissipative Media', *Proc. ASCE J. Eng. Mech. Div.,* Vol. 86, 1960, pp. 77–105.

[15] CLIFTON, R.J. and BODNER, S.R., 'An Analysis of Longitudinal Elastic-Plastic Pulse Propagation', *J. Appl. Mech.,* Vol. 31, 1966, pp. 248–255.

[16] TUSCHAK, P.A. and SCHULTZ, A.B., 'Determination of the Unloading Boundary in Longitudinal Elastic-Plastic Stress Wave Propagation', *J. Appl. Mech.,* Vol. 38, 1971, pp. 888–898.

[17] VON KARMAN, Th., BOHNENBLUST, H.F. and HYERS, D.H., 'The Propagation of Plastic Waves in Tension Specimens of Finite Length', NDRC Report No. A-103, OSRD No. 946, 1943.

[18] CLIFTON, R.J. and TING, T.C.T., 'The Elastic-Plastic Boundary in One-Dimensional Wave Propagation', *J. Appl. Mech.,* Vol. 35, 1968, pp. 812–814.

[19] YU, J.L., WANG, L.L. and ZHU, Z.X., 'Basic Properties of Elastic-Plastic Boundaries in Stress Wave Propagation in a Bar', *Acta Mechanica Solida Sinica,* Aug. 1982, No. 3, pp. 313–323 (in Chinese).

[20] YU, J.L., WANG, L.L. and ZHU, Z.X., 'Determination of Propagation Velocity of Elastic-Plastic Boundaries in a Bar', *Acta Mechanica Solida Sinica,* March 1984, No. 1, pp. 16–26 (in Chinese).

[21] WANG, L.L., ZHU, Z.X. and YU, J.L., 'On Discontinuous Properties of Elastic-Plastic Boundary in Elastic-Plastic Plane Wave Propagation', *Explosion and Shock Waves,* Vol. 3, No. 1, 1983, 1–8 (in Chinese).

[22] TING, T.C.T., 'On the Initial Speed of Elastic-Plastic Boundaries in Longitudinal Wave

Propagation in a Rod', *J. Appl. Mech.,* Vol. 38, 1971, pp. 441–447.

[23] CLIFTON, R.J., 'Elastic-Plastic Boundaries in Combined Longitudinal and Torsional Plastic Wave Propagation', *J. Appl. Mech.,* Vol. 35, 1968, pp. 782–786.

[24] TING, T.C.T., 'On the Initial Slope of Elastic-Plastic Boundaries in Combined Longitudinal and Torsional Wave Propagation', *J. Appl. Mech.,* Vol. 36, 1969, pp. 203–211.

[25] TING, T.C.T., 'A Unified Theory on Elastic-Plastic Wave Propagation of Combined Stress', *Foundations of Plasticity,* ed. by A. Sawezuk, Noordhoff Int. Publ., Leyden, 1973, pp. 301–316.

[26] MANDEL, J., 'Ondes Plastique dans un Milieu Indefini a Trois Dimensions', *J. Mécanique,* Vol. 1, 1962, pp. 3–30.

[27] BALABAN, M.M., Green, A.E. and NAGHDI, P.M., 'Acceleration Waves in Elastic-Plastic Materials', *Int. J. Eng. Sci.,* Vol. 8, 1970, pp. 315–335.

[28] TING, T.C.T., 'Shock Waves and Weak Discontinuities in Anisotropic Elastic-Plastic Media', *Propagation of Shock Waves in Solids,* ed. by E. Varley, ASME, AMD Vol. 17, 1976.

[29] RANIECKI, B., 'Ordinary Waves in Inviscid Plastic Media', *Mechanical Waves in Solids,* ed. by J. Mandel and L. Brun, CISM Courses and Lectures No. 222, Springer-Verlag, NY, 1976, pp. 157–219.

A.M. Skudra and Yu.M. Tarnopol'skii

7

Engineering mechanics of composites

7.1 Introduction

The contribution of Yu.N. Rabotnov to the mechanics of composites is extensive; only the essential ones will be singled out. Advanced fibrous composites are heterogeneous anisotropic. Composite behavior depends on the matrix which can be polymeric, metallic or ceramic, and the angle of load application relative to the direction of reinforcement. Heterogeneity can occur at the micro-micro-heterogeneity level in a lamina consisting of fibers and matrix and at the macro-level for a laminated structure. The variety of possible combinations of deformation and failure modes in composites make the study of this class of materials challenging as it can encompass the different disciplines in solid mechanics.

The central problem concerning the mechanics of fibrous composites, as identified in [1], is the influence of reinforcement on structure behavior. Such an approach is outside the scope of the theory of solid mechanics. Composites are tailor-made so that the properties are governed by the fiber layup and matrix which can be adjusted to optimize the performance.

It is the tailor-made feature of composites that can optimize the use of such materials. Extensive studies have been reported in [2 – 4] while a number of other references can be found in [5] published by Mir Publishing House in the series on *Advances of Science and Techniques.* The fundamental ideas can also be extended to manufacture viscoelastic polymers and ductile metal matrices.

Composites, being heterogeneous, anisotropic and inelastic, require special consideration when modelling their behavior. Difficulties, however, are encountered by the conventional methods. Engineering models [6, 7] are thus recommended to assess the influence of the constituents on the overall composite behavior.

If the material in each lamina is homogeneous across the thickness layers, say 0°; $\pm 90^o$; $0/90^o$, then heterogeneity can be neglected and the fiber composite can be modelled as an anisotropic medium with reinforcements [8 – 10]. It is analogous to ribbed plates and shells where the influence of stringers and frames can be smeared and represented by anisotropic behavior [11]. The structural behavior of composites cannot be overemphasized. A technique for smoothing the energy has been introduced in [8] in such a way that the theory of anisotropic elasticity [12, 13] can be utilized.

Modelling of a heterogeneous composite by a homogeneous and anisotropic body has been used to analyze the vibration of beams, plates and shells. Instead of using the Kirchhoff equivalent shear, the influence of the transverse tensile strengths of composites has been studied [14, 15]. The treatment of a composite as an anisotropic body can be found in [16, 17]. The shear strength of fiber reinforced composites, especially the interface, has been investigated in relation to bars [18], plates [8, 14], shells [10, 15], and reinforced structures [19]; when the fiber reinforcement and principal stresses do not coincide, the consideration of viscoelasticity becomes necessary. Refer to the work reported in [20] associated with the hereditary theory. As the polymeric matrix is sensitive to changes in mechanical forces and temperature history, the fabrication process can alter composite behavior; this is more sensitive for composites than metallic materials [21]. In particular, interlaminar shear and transverse tensile strengths can be greatly affected. The optimum condition of molding has been emphasized [22, 23].

At present, separate theories are advanced when the composite matrix is altered; the three common constituents are polymer, metal and ceramic, each being analyzed and fabricated differently. In this chapter we shall focus attention on the concept of Rabotnov dealing with creep and stress-rupture of reinforced plastics. Use will be made of a function that weights the damage. The work consists of the research done by the author and his co-workers.

Creep in reinforced plastics takes place mainly in the matrix. A singular creep kernel in the form of $з_\alpha$ has been proposed in [24] which is used in the theory of linear heredity of Volterra. It can account for the rheological properties of the material. Attention was also given to composite failure by delamination and fiber cracking, and brittle fracture under sustained load. The latter process involves crack growth followed by the formation of a through crack leading to fracture. A damageability parameter w is introduced for evaluating the crack growth process.

7.2 Visco-elastic properties of constituents

Deformation of the polymeric binder in uniaxial loading can be described by an integral equation

$$\varepsilon(t) = \frac{1}{E_A}[\sigma(t) + \int_0^t K_A(t - \theta)\sigma(\theta)\, d\theta]. \tag{7.1}$$

The function $K_A(t - \theta)$ is the kernel determined by experiment. Under constant stress, i.e., $\sigma(t)$ = const, Equation (7.1) becomes

$$\varepsilon(t) = \frac{\sigma}{E_A}[1 + \int_0^t K_A(t - \theta)\, d\theta] = D(t)\sigma. \tag{7.2}$$

The creep function $D(t)$ can be preassigned in the form of graphs, tables and analytical relationships.

The experimental results show that the strain rate under constant stress tends to infinity initially. The creep kernel at $t = 0$ should therefore be singular. This singularity should be weak so that the initial strains remain finite. To this end, the Duffing kernel [26] is introduced; the creep curve, corresponding to Equation (7.2), increases without bound, i.e., the modulus of long-term elasticity to zero. In [24] is proposed the fractional-exponential

$$\mathfrak{Z}_{\alpha A}(-\beta_A, t-\theta) = (t-\theta)^{\alpha} \sum_{n=0}^{\infty} \frac{-\beta^n (t-\theta)^{n(1+\alpha)}}{\Gamma[(n+1)(1+\alpha)]}.$$

It follows that the creep function of the polymeric matrix takes the form

$$D(t) = \frac{1}{E_A}[1 + \lambda_A \int_0^t \mathfrak{Z}_{\alpha A}(-\beta_A, t-\theta)\mathrm{d}\theta]. \tag{7.3}$$

An application of the function $\mathfrak{Z}_{\alpha A}$ is related to the determination of the parameters E_A, α_A, β_A and λ_A, the value of which are assumed to be the rheological properties of the material [27, 28]. If α_A, β_A and λ_A are known, then the creep function $D(t)$ can be determined from Equation (7.3). Expanding $\mathfrak{Z}_{\alpha A}$ as a power series, special charts in [29] can be used to solve Equation (7.3) numerically. The following approximation [30] is obtained:

$$\int_0^t \mathfrak{Z}_{\alpha A}(-\beta_A, t-\theta)\mathrm{d}\theta \approx \frac{1}{\beta_A}[1 - \exp(-\beta_A \nu t^{1+\alpha_A})] \tag{7.4}$$

where

$$1 < \alpha_A < 0; \; \nu = (1+\alpha_A)^{1+\alpha_A}.$$

Making use of Equations (7.4) and (7.2), there results

$$\varepsilon_A(t) = \varepsilon_0\left\{1 + \frac{\lambda_A}{\beta_A}[1 - \exp(-\beta_A \nu t^{1+\alpha_A})]\right\} \tag{7.5}$$

where ε_0 is the elastic strain. As $t \to \infty$, Equation (7.5) gives

$$\varepsilon_A(\infty) = \varepsilon_0\left(1 + \frac{\lambda_A}{\beta_A}\right).$$

Hence, if the strains under instantaneous and infinitely sustained load are known, the relation

$$\frac{\lambda_A}{\beta_A} = \frac{\varepsilon_A(\infty)}{\varepsilon_0} - 1 \tag{7.6}$$

is obtained. The parameters α_A and β_A are given by

$$\alpha_A = -1 + \frac{1}{\ln t_2/t_1} \ln \frac{\ln(1 - \frac{\beta_A}{\lambda_A}a_1)}{\ln(1 - \frac{\beta_A}{\lambda_A}a_2)}$$

$$\beta_A = -\frac{\ln(1 - \frac{\beta_A}{\lambda_A}a_1)}{[(1+\alpha_A)t_1]^{(1+\alpha_A)}} \tag{7.7}$$

where

$$a_1 = \frac{\varepsilon(t_1)}{\varepsilon_0} - 1 \quad a_2 = \frac{\varepsilon(t_2)}{\varepsilon_0} - 1.$$

Here, t_1 and t_2 are two independent fixed values of loading time. According to Equations (7.7), the parameters α_A and β_A are not unique. The method for determining the creep parameters α_A, β_A and λ_A by using Equations (7.6) and (7.7) is nevertheless useful. The experimental data for the creep of a phenolformaldehyde matrix under compression is presented in Figure 7.1. The approximate curve is constructed from Equation (7.5) for $\alpha_A = -0.35$, $\beta_A = 0.21\ \text{days}^{0.65}$ and $\lambda_A = 0.17\ \text{days}^{-0.65}$. Under these considerations, the creep function for the polymeric matrix in Equations (7.3) and (7.4) takes the form

$$D(t) = \frac{1}{E_A}\left\{1 + \frac{\lambda_A}{\beta_A}[1 - \exp(-\beta_A \gamma t^{1+\alpha_A})]\right\}. \tag{7.8}$$

Creep kernels approximated by functions in the form of a sum of exponents have been widely used. The number of independent parameters, however, tends to increase with the number of terms in the sum. If the creep kernel contains one exponent; it becomes a particular case of Equation (7.4) with $t = 0$.

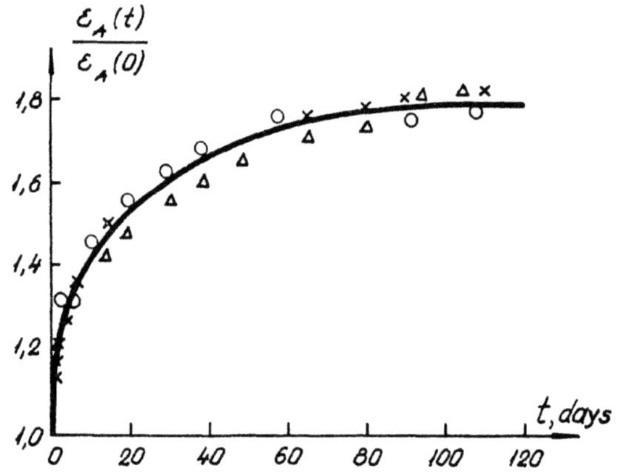

Fig. 7.1: Creep of phenolformaldehyde matrix under compression.

Equation (7.2) defines the creep of a polymeric matrix under uniaxial state of stress. The stress state in a polymeric matrix in reinforced plastics, however, is multiaxial regardless of loading. The deformation of a polymeric matrix under triaxial stress can be found by assuming that there is no change in volume with time [25]:

$$\varepsilon_1(t) = D(t)\left[\sigma_1 - \frac{\sigma_2 + \sigma_3}{2}\right] + \frac{1 - 2\nu}{E_A}(\sigma_2 + \sigma_3) \tag{7.9}$$

where σ_1, σ_2, σ_3 are the stress components, ν_A and E_A are respectively Poisson's ratio and the elastic modulus of the polymeric matrix, and $D(t)$ is the creep function under uniaxial stress.

Similarly, the expressions for strains $\varepsilon_2(t)$ and $\varepsilon_3(t)$ may also be written.

If the creep equation of the polymeric matrix in shear is symbolically written as

$$\gamma(t) = J(t)\tau \tag{7.10}$$

then the condition of elastic compressibility gives the creep function

$$J(t) = 3D(t) - \frac{1 - 2\nu_A}{E_A} \tag{7.11}$$

for uniaxial load.

The viscoelastic properties of the polymeric matrix for a linear stress-strain relation are characterized by the time function, $D(t)$, which can be obtained by experimental considerations. If the creep curve of the polymeric matrix in shear is obtained experimentally, it is easy to establish the function $J(t)$, and then the function $D(t)$ by means of Equations (7.11).

Nonlinearity in creep can occur at elevated temperatures and/or high loads. Such behavior can also occur at room temperature. Presented in Figure 7.2 are creep curves of the matrix ЗДТ – 10 in uniaxial tension and compression under isothermal conditions ($T = 22^\circ$C). The degree of nonlinearity changes with time. Similarity of creep curves or similarity of isochronous curves are not observed. Creep curves for the polymeric matrix ЗДТ – 10 under axial loading, can be assessed from a rheological equation of the form

$$\varepsilon(t) = \frac{1}{E_A}\left\{\sigma_A(t) + k_A\sigma_A^3(t) + \int_0^t k_A(t - \theta)[\sigma_A(\theta) + \right.$$

$$\left. + (d_A + (k_A - d_A)e^{-\kappa_A\theta})\sigma_A^3(\theta)]\, d\theta\right\}. \tag{7.12}$$

The kernel $k_A(t - \theta)$ characterizes creep in the linear strain range at low load level while the three parameters k_A, d_A and κ_A account for the nonlinear behavior of the matrix. In some cases, the kernel takes the form

$$k_A(t - \theta) = \sum_{j=1}^{n} C_{Aj} \exp[-\alpha_{Aj}(t - \theta)].$$

Accuracy can be improved by taking more terms. In practice, it suffices to take $n = 2$ or $n = 3$. If $\sigma_A = $ const, Equation (7.12) becomes

$$\varepsilon_A(t) = \frac{1}{E_A}\left\{\sigma_A + k_A\sigma_A^3 + \sum_{j=1}^{n}\left[\frac{C_{Aj}}{\alpha_{Aj}}(1 - e^{-\alpha_{Aj}t})(\sigma_A + d_A\sigma_A^3) + \right.\right.$$

$$\left.\left. + \frac{C_{Aj}}{\alpha_{Aj} - \kappa_A}(e^{-\kappa_A t} - e^{-\alpha_{Aj}t})(k_A - d_A)\sigma_A^3\right]\right\}. \tag{7.13}$$

Nonlinear creep strain can be described by Equation (7.13) with $4 + 2n$ constants. The

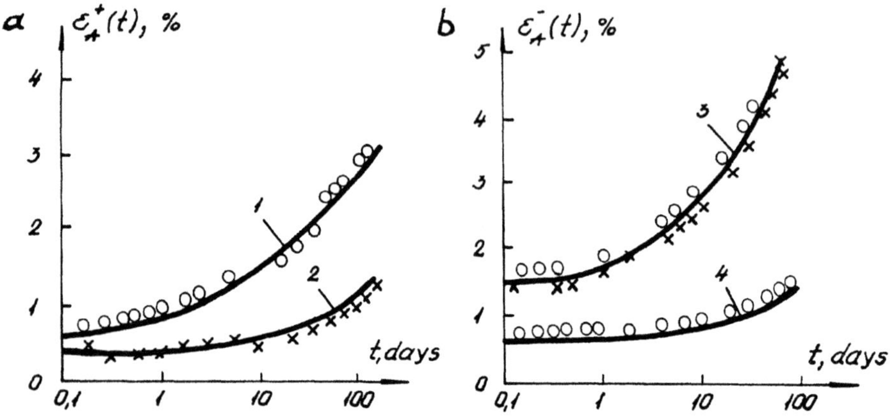

Fig. 7.2: Creep of an epoxy matrix; (a) tension and (b) compression for σ_A = MPa, 200 MPa(1), 120 MPa(2), 500 MPa(3) and 240 MPa(4).

$1 + 2n$ constants describe linear creep; the remaining three describe the nonlinear behavior. Approximation by the method of least squares is used to obtain parameters C_{Aj} and α_{Aj} in the kernel for the linear behavior. The epoxy matrix ЗДТ – 10 has the properties $C_{A1} = 0.57$, $C_{A2} = 0.028$, $C_{A3} = 0.01056$, $\alpha_{A1} = 2$ days^{-1}, $\alpha_{A2} = 0.60833$ days^{-1}, and $\alpha_{A3} = 0.007143$ days^{-1} as in Figure 7.2(a). Nonlinear creep requires the determination of the parameters k_A, d_A and β_A. When $t = 0$, Equation (7.13) takes the form

$$\varepsilon_A(t) = \frac{1}{E_A}(\sigma_A + k_A\sigma_A^3). \tag{7.14}$$

The parameters k_A is obtained by approximating the nonlinear strain curve for short-time loading. Hence Equation (7.14) gives

$$k_A = \frac{E_A\varepsilon_A(0) - \sigma_A}{\sigma_A^3}. \tag{7.15}$$

Similarly, if $t \to \infty$, Equation (7.13) can be used to yield

$$d_A = \frac{aE_A\varepsilon_A(\infty) - (1 + a)\sigma_A}{\sigma_A^3} - ak_A \tag{7.16}$$

where

$$a = \left(\sum_{j=1}^{n} \frac{C_{aj}}{\alpha_{Aj}}\right)^{-1}$$

and $\varepsilon_A(\infty)$ is the steady creep strain corresponding to σ_A. The parameter κ_A determined from Equation (7.13) for $t = t_1$, characterizes the nonlinear time behavior of the matrix. Making use of Equations (7.15) and (7.16) and data in Figure 7.2, the parameters $k_A = 0.35 \times 10^{-3}$ (MPa)$^{-2}$, $d_A^+ = 0.54 \times 10^{-2}$ (MPA)$^{-2}$ and $\kappa_A^+ =$ 0.4 days^{-1} are obtained for tensile loading, $k_A^- = 0.104 \times 10^{-4}$ (MPa)$^{-2}$, d_A^-

$= 0.61 \times 10^{-3}$ (MPa)$^{-2}$, and $\kappa_A^- = 0.55$ days^{-1} for compressive loading for describing the nonlinear behavior of the epoxy matrix ЗДТ – 10. There is good agreement between experimental data and design creep curves constructed from Equation (7.13).

The linear-elastic and linear-viscoelastic properties of the epoxy matrix ЗДТ – 10 in tension and compression are about the same; the nonlinear properties however, are more pronounced in tension. Equation (7.13) provides sufficient accuracy for describing the creep behavior of the polymeric matrix subjected to uniaxial tension, compression or shear.

The above method also applied to fiber reinforced composites. Unlike the matrix material, fibers have relatively small creep; they are practically independent of loading rate. Therefore, boron, carbon or glass fibers can be considered as elastic. For organic fibers, this cannot be assumed. According to the results in [31], Kevlar–49 fibers do creep. As shown in Figure 7.3, creep also prevails in high-strength filaments and microplastics which are made of threads impregnated with an epoxy resin and then heat treated. The specific creep defined as the strain-to-initial strain ratio in Figure 7.3 are averages; they were constructed from the results of long-time tests[32] at stress levels up to 0.6 of the breaking stress under short-term loading. Within the range of applied stresses, the stress-strain relation is linear at any instant of loading. Organic fibers and microplastics can thus be modelled by the linear theory of viscoelasticity and the creep curves may be obtained from Equation (7.2).

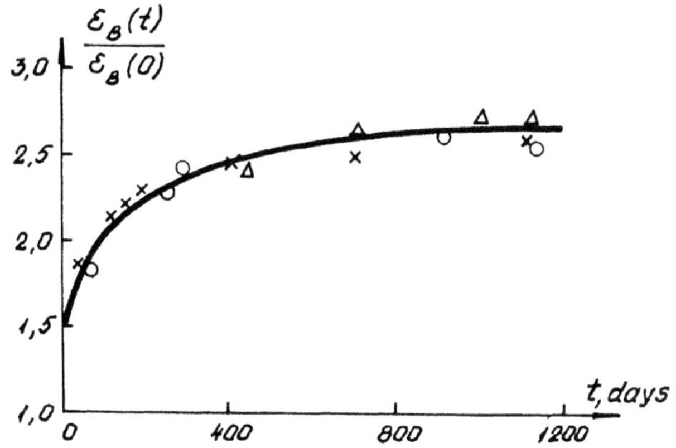

Fig. 7.3: Creep of organic fiber filament in tension.

7.3 Creep of unidirectionally reinforced plastic

Axial creep along fiber direction

Creep along the fibers of unidirectionally reinforced composites can be described by the viscoelastic properties of the constituents. Determination of creep deformation can be reduced to solving a system of two linear integral equations of the type in Equation (7.1) that account for compatibility and equilibrium. Use was made of the fractional-exponential and exponential functions as kernels for describing the creep of the matrix and fibers [33, 37]. Only the simplest case will be presented where the creep behavior of the matrix and fibers will be approximated by exponential functions. For axial tension, the relations

$$\varepsilon_A(t) = \frac{1}{E_A}\left[\sigma_A(t) + \lambda_A \int_0^t e^{-\beta_A(t-\theta)}\sigma_A(\theta)\,d\theta\right] \quad (7.17)$$

$$\varepsilon_B(t) = \frac{1}{E_A}\left[\sigma_B(t) = \lambda_B \int_0^t e^{-\beta_A(t-\theta)}\sigma_B(\theta)\,d\theta\right] \quad (7.18)$$

are obtained. Equilibrium requires the stress $\sigma_A(t)$ in the matrix and $\sigma_B(t)$ in the fiber be related as

$$(1-\psi)\sigma_A(t) + \psi\sigma_B(t) = \langle \sigma_{11} \rangle \quad (7.19)$$

where $\langle \sigma_{11} \rangle$ is constant mean stress of a unidirectional reinforced layer. Without going into detail, solutions of Equations (7.17) – (7.19) will be given for $\langle \varepsilon_{11}(t) \rangle = \varepsilon_A(t) = \varepsilon_B(t)$: let us present here only the final result obtained in our work

$$\langle \varepsilon_{11}(t) \rangle = s_{11}\left(1 + \frac{a_3}{b_3}\right)[1 + A_1 e^{-z_1 t} + A_2 e^{-z_2 t}]\langle \sigma_{11} \rangle \quad (7.20)$$

where

$$A_1 = \frac{a_3 - z_1(a_1 + a_2)}{(z_1 - z_2)(b_3 + a_3)} z_2; \quad A_2 = \frac{a_3 - z_2(a_1 + a_2)}{(z_2 - z_1)(b_3 + a_3)} z_1$$

$$z_{1,2} = \frac{b_1 + b_2}{2}\left(1 \mp \sqrt{1 - 4\frac{b_3}{(b_1 + b_2)^2}}\right)$$

$$a_1 = \lambda_B[1 - (1-\psi)E_A s_{11}]; \quad a_2 = \lambda_A[1 - \psi\acute{E}_{Bz} s_{11}]$$

$$a_3 = \lambda_A\lambda_B + \beta_A a_1 + \beta_B a_2$$

$$b_1 = \beta_B + (1-\psi)\lambda_B E_A s_{11}; \quad b_2 = \beta_A + \psi\lambda_A E_{Bz} s_{11}$$

$$s_{11} = \frac{1}{E_{11}}; \quad b_3 = b_1\beta_A + b_2\beta_B - \beta_A\beta_B.$$

Here, E_{11} is the elastic modulus of a unidirectional reinforced plastic in the direction of the fiber. The creep curve, obtained from Equation (7.20) for an organic plastic is shown in Figure 7.4. For $t \to \infty$, Equation (7.20) yields

$$\langle \varepsilon_{11}(\infty) \rangle = s_{11}\left(1 + \frac{a_3}{b_3}\right)\langle \sigma_{11} \rangle. \quad (7.21)$$

The curves are constructed according to Equation (7.20) and based on initial data in [49]: $\alpha_A = 4 \times 10^{-3}$ (days)$^{-1}$, $c_A = 17.6 \times 10^{-3}$ (days)$^{-1}$, $\alpha_B = 9.5 \times 10^{-3}$ (days)$^{-1}$,

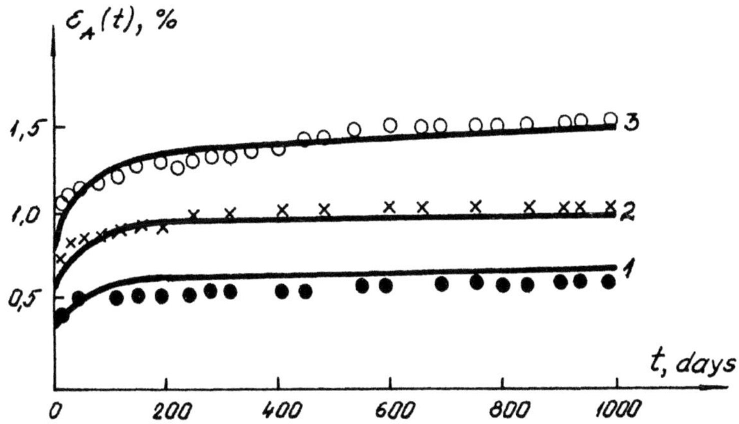

Fig. 7.4: Creep of unidirectional reinforced organic plastic for $\langle\sigma_{11}\rangle$ = 300 MPa(1), $\langle\sigma_{11}\rangle$ = 450 MPa(2) and $\langle\sigma_{11}\rangle$ = 600 MPa(3).

$c_B = 11.9 \times 10^{-3}$ (days)$^{-1}$, $E_A = 4.5 \times 10^3$ MPa; $E_{Bz} = 1.4 \times 10^5$ MPa and $\psi = 0.65$. Once $\langle \varepsilon_{11}(t) \rangle$ is known the time dependent stresses can be found:

$$\sigma_A(t) = E_A \left[\langle \varepsilon_{11}(t) \rangle - \lambda_A \int_0^t e^{-(\beta_A + \lambda_A)(t-\theta)} \langle \varepsilon_{11}(\theta) \rangle \, d\theta \right] \tag{7.22}$$

$$\sigma_B(t) = E_{Bz} \left[\langle \varepsilon_{11}(t) \rangle - \lambda_B \int_0^t e^{-(\beta_B + \lambda_B)(t-\theta)} \langle \varepsilon_{11}(\theta) \rangle \, d\theta \right]. \tag{7.23}$$

Substituting the expression $\langle \varepsilon_{11}(t) \rangle$ in Equations (7.20) and (7.23), it is found that

$$\sigma_A(t) = E_A s_{11} \left(1 + \frac{a_3}{b_3} \right) \left[\lambda_A \left(\frac{1}{\lambda_A + \beta_A} + \frac{A_1}{\lambda_A + \beta_A - z_1} + \right. \right.$$
$$\left. + \frac{A_2}{\lambda_A + \beta_A - z_2} \right) e^{-(\beta_A + \lambda_A)t} + \frac{A_1(\beta_A - z_1)}{\lambda_A + \beta_A - z_1} e^{-z_1 t} +$$
$$\left. + \frac{A_2(\beta_A - z_2)}{\lambda_A + \beta_A - z_2} e^{-z_2 t} + \frac{\beta_A}{\lambda_A + \beta_A} \right] \langle \sigma_{11} \rangle$$

$$\sigma_B(t) = E_{Bz} s_{11} \left(1 + \frac{a_3}{b_3} \right) \left[\lambda_B \left(\frac{1}{\lambda_B + \beta_B} + \frac{A_1}{\lambda_B + \beta_B - z_1} + \right. \right. \tag{7.24}$$
$$+ \left. \frac{A_2}{\lambda_B + \beta_B - z_2} \right) e^{-(\beta_B + \lambda_B)t} + \frac{A_1(\beta_B - z_1)}{\lambda_A + \beta_A - z_1} e^{-z_1 t} +$$

$$+ \frac{A_2(\beta_B - z_2)}{\lambda_B + \beta_B - z_2} e^{-z_2 t} + \frac{\beta_B}{\lambda_B + \beta_B} \Bigg] \langle \sigma_{11} \rangle$$

in which a_3, b_3, A_1, A_2, z_1 and z_2 can be determined from Equations (7.20). Equation (7.24) shows that as the stresses relax in the matrix, the stresses in the fibers increase.

Creep in longitudinal shear

Figure 7.5 shows the longitudinal shear of a viscoelastic body reinforced unidirectionally; an infinite number of layers with infinitely small thickness is assumed to be parallel to the plane of loading. It is assumed that each layer is under uniform stress and the average strains of all layers at a given instant are equal. The shear strain in the layers is the sum of all strains in the matrix and fibers. Recall that the matrix is viscoelastic and the fibers are elastic. The stress-strain state in the reinforced layer is reduced to solving the system of equations:

$$\begin{aligned}
&(1 - \psi_i)\nu_{Ai}(t) + \psi_i \nu_{Bi}(t) = \langle \nu_{11\perp}(t) \rangle \\
&G_A \nu_{Ai}(t) = \tau_{Ai}(t) + \int_0^t k_\tau(t - \theta)\tau_{Ai}(\theta)\, d\theta \\
&G_{Br_z} \nu_{Bi}(t) = \tau_{Bi}(t) \qquad (7.25) \\
&\tau_{Ai}(t) = \tau_{Bi}(t) = \tau_{11\perp i}(t) \\
&\langle \tau_{11\perp}(t) \rangle = \frac{r_B}{p} \int_0^{\pi/2} \tau_{11\perp i}(\phi) \cos\phi \, d\phi + \left(1 - \frac{r_B}{p}\right) \tau_A(t) \\
&\tau_A(t) = G_A \left[\langle \nu_{11\perp}(t) \rangle - \int_0^t R_\tau(t - \theta) \langle \nu_{11\perp}(\theta) \rangle \, d\theta \right]
\end{aligned}$$

where $\tau_A(t)$ is the stress in those layers with no fiber and $R_\tau(t - \theta)$ are the creep and relaxation kernels for the matrix in shear. If $\langle \nu_{11\perp} \rangle$ = const., a solution of Equation (7.25) can be found that gives the average stress in a unidirectional reinforced layer and shear stresses in the components.

Suppose that $\langle \varepsilon_{11\perp} \rangle$ = const., then a redistribution of stresses occurs; the shear stresses at $t = 0$ and $t = \alpha$ will be different. Figure 7.6 shows that the shear stress on the plane at M has not changed, which is characteristic of a particular plastic with $\tau_{11\perp m}$ being constant. The location of the plane M is defined by the angle

$$\phi_m = \arccos \frac{l}{r_B} \left[1 + \frac{1 - \eta_{11\perp}}{\eta_{11\perp} - \eta_{\tau A}} \frac{G_A}{G_{Brz}} \right]^{-1} \qquad (7.26)$$

where $\eta_{\tau A}$ determines the extent of creep in the matrix under shear; it is defined by the relation between creep functions in shear $J(t)$ at $t \to \infty$ and $t \to 0$. Taking Equation (7.11) into account the result is

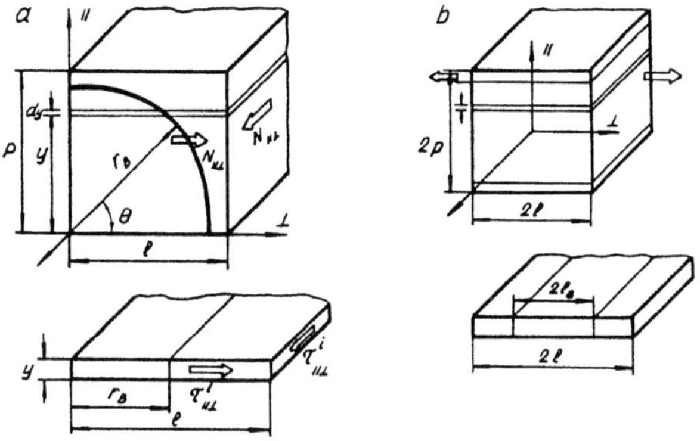

Fig. 7.5: Structural element of unidirectional reinforced plastic; (a) longitudinal shear and (b) transverse load.

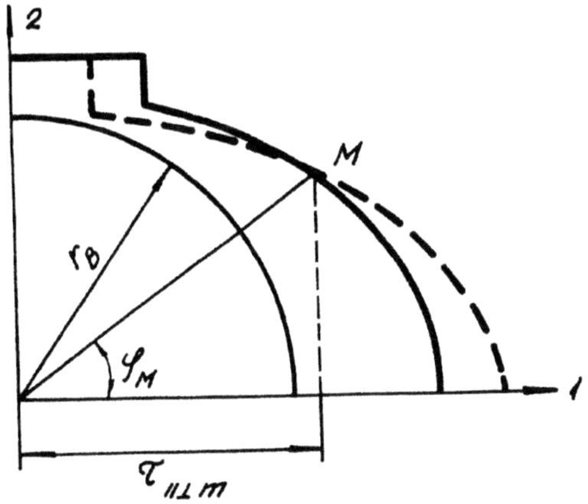

Fig. 7.6: Shear stress distribution in reinforced plastic upon load application (solid curve) and after constant sustained load (dotted line).

$$\eta_{\tau A} = \frac{J(\infty)}{J(0)} = 1 + \frac{3}{2(1+\nu_A)}\frac{\lambda_A}{\beta_A}. \tag{7.27}$$

The mean strain of the Mth layer is calculated as a sum of the strains in the fibers and matrix; it is equal to the mean strain in the plastic:

$$\nu_{11\perp}m = \langle \nu_{11\perp} \rangle = \frac{r_B}{l}\cos\phi_M \nu_{BM} + \left(1 - \frac{r_B}{l}\cos\phi_M\right)\nu_{AM}. \tag{7.28}$$

The stress in the Mth layer is determined from Equation (7.28):

$$\tau_{11\perp m} = \frac{G_{Brz}}{G_{11\perp}} \left[\frac{G_{Brz}}{G_A} \left(1 - \frac{r_B}{l} \cos \phi_M \right) + \frac{r_B}{l} \cos \phi_M \right]^{-1} \langle \tau_{11\perp} \rangle . \tag{7.29}$$

Substitution of Equation (7.26) into (7.29) gives

$$\tau_{11\perp m} = \frac{G_{Brz}}{G_{11\perp}} \left[1 - \frac{1 - \eta_{11\perp}}{1 - \eta_{\tau A}} \left(1 - \frac{G_A}{G_{Brz}} \right) \right] \langle \tau_{11\perp} \rangle \tag{7.30}$$

where $G_{11\perp}$ is the shear modulus of a unidirectional reinforced plastic.

The creep behavior of a unidirectional reinforced plastic under longitudinal shear can be obtained from a generalization of Equation (7.28) for sustained constant stress $\tau_{11\perp m}$. The dependence of mean shear strain on load duration is given by

$$\langle \nu_{11\perp}(t) \rangle = s_{66}(t) \langle \tau_{11\perp} \rangle . \tag{7.31}$$

Here, $s(t)$ is the creep function that can be found by

$$s_{66}(t) = g_1 + g_2 J(t) \tag{7.32}$$

where

$$g_1 = \frac{\eta_{11\perp} - \eta_{\tau A}}{1 - \eta_{\tau A}} \cdot \frac{1}{G_{11\perp}}; \quad g_2 = \frac{1 - \eta_{11\perp}}{1 - \eta_{\tau A}} \cdot \frac{G_A}{G_{11\perp}}.$$

Equations (7.31) and (7.32) show that the viscoelastic properties of the matrix determine the composite behavior. The solid line in Figure 7.7 represents the creep curve for the phenolformaldehyde fiberglass plastic in longitudinal shear. The corresponding data are $\psi = 0.52$, $\sigma_{Brz} = 2.8 \times 10^4$, $E_A = 0.36 \times 10^4$ MPa, $\nu_A = 0.35$, $\alpha_A = -0.35$, $\beta_A = 0.21$ days$^{-0.65}$, and $\lambda_A = 0.17$ days$^{-0.65}$. The experimental results obtained at shear stress $\langle \tau_{11\perp} \rangle = 22.5$ MPa are also given. The assumptions are validated by the good agreement between the experimental data and calculated results.

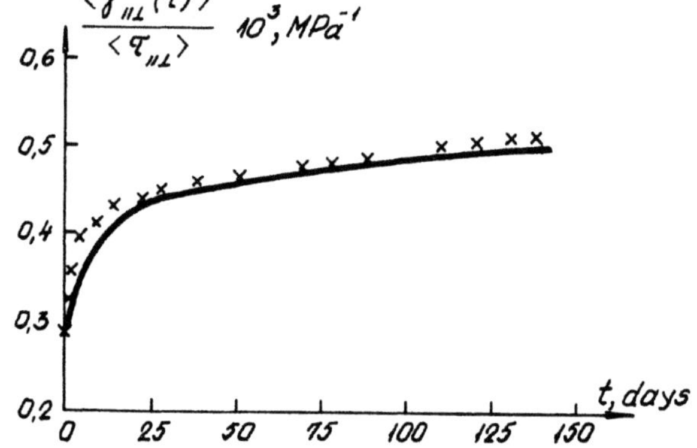

Fig. 7.7: Deformation of a phenolformaldehyde fiberglass plastic in longitudinal shear.

Creep in transverse loading

In order to find the stress-strain state of the structural elements of a unidirectional reinforced plastic under sustained transverse loading, a volumetric boundary-value problem for the heterogeneous binary medium must be solved. No method is presently available for obtaining an accurate solution. The unidirectionally reinforced plastic is modelled as a doubly-periodic structure [35] which is shown in Figure 7.5(b). By dividing the element into infinitely thin layers, a system of equations was derived for determining the stress-strain state in the entire reinforced plastic. The fibers are assumed to be transversally isotropic and the matrix behaves in accordance with Equation (7.1). Relations are thus obtained [35, 36] for determining the stresses in the fibers and matrix under a constant sustained load applied in the transverse direction of a uniaxially reinforced plastic. One of the conclusions [36] is that the matrix is under nonuniform triaxial stress. For reinforced plastics, such as fiberglass and boron plastics, this stress does not depend on duration of the loading. The matrix can thus be assumed to be in constant stress for the transverse creep of fiberglass and boron plastics. Consequently, Equation (7.9) may be used. Referring to Figure 7.5(b), the creep of reinforced plastic in the direction of loading is equal to that in any other layers.

Consider a layer having a maximum volume fraction of fibers, i.e., the layer, for which $\phi = 0$ in Figure 7.5(b). The strain in the layer is equal to

$$\langle \varepsilon_{\perp}(t) \rangle = \frac{r_B}{l}\varepsilon_{Br} + \left(1 - \frac{r_B}{l}\varepsilon_A(t)\right) \tag{7.33}$$

where ε_{Br} is the elastic strain of the fiber in the direction of the mean stress $\langle \sigma_{\perp} \rangle$ and is determined from Hooke's law. The strain $\varepsilon_A(t)$ in the matrix is determined from Equation (7.9). The creep strain of a unidirectional reinforced plastic under transverse loading is determined by

$$\langle \varepsilon_{\perp}(t) \rangle s_{22}(t) \langle \sigma_{\perp} \rangle \tag{7.34}$$

where $s_{22}(t)$ is the creep function of the plastic in the transversal direction. The creep function is [36]

$$s_{22}(t) = d_1 + d_2 D(t) \tag{7.35}$$

where

$$d_1 = \frac{r_B}{l} - \frac{\sigma_r}{E_{Br}} + \left(1 - \frac{r_B}{l}\right)\frac{1 - 2\nu_A}{2E_A}(\overline{\sigma_\theta} - \overline{\sigma_z})$$

$$d_2 = \left(1 - \frac{r_B}{l}\right)\left(\overline{\sigma_r} - \frac{\overline{\sigma_\theta} + \overline{\sigma_z}}{2}\right). \tag{7.36}$$

Here, $\overline{\sigma_r}$, $\overline{\sigma_\theta}$ and $\overline{\sigma_z}$ refer to the stress state in the constituents of the reinforced plastic as a function of the elastic properties, the volume fraction of fibers and geometry of the layups. Their determination can be found in [34]. Figure 7.8 gives the creep data for a unidirectional reinforced fiberglass in a phenolformaldehyde matrix with a volume fraction of $\psi = 0.52$ as obtained from Equation (7.34). A rectangular fiber distribution in the transverse section of the plastic was assumed. Use was made of the following data: $E_{Bz} = E_{Br} = 7 \times 10^4$ MPa, $\nu_{Bzr} = \nu_{Br\theta} = 0.22$, $E_A = 0.36 \times 10^4$ MPa, $\nu_A = 0.35$, $\alpha_A = -0.35$, $\beta_A = 0.21$ days$^{-0.65}$, $\lambda_A = 0.17$ days$^{-0.65}$ and $\langle \sigma_\perp \rangle = 1$ MPa. A compressive stress $\langle \sigma_\perp \rangle = 45$ MPa was used. The transverse creep of anisotropic fibers such as carbon and organic material can only be treated approximately. In this case, the stress state varies with time and the maximum stress in the matrix of a carbon plastic is about 30% of the initial value, when both the fibers and matrix are viscoelastic. Time variation of the volumetric stress state needs to be considered [37]. As an approximation, assume a constant maximum stress in the nth layer which can be repeated to yield the behavior of the entire composite. The location of this layer is defined by

$$\phi_N = \arccos \frac{l}{r_B} \left[1 + \frac{1 - \eta_\perp}{1 - \eta_A} \cdot \frac{E_A}{E_{Br}} \right]^{-1} \tag{7.37}$$

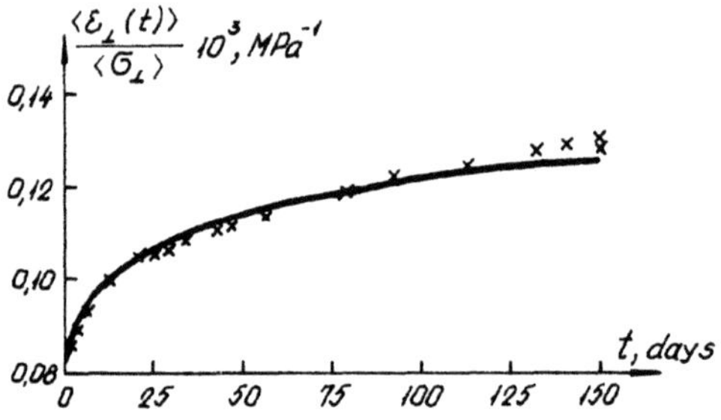

Fig. 7.8: Deformation of a phenolformaldehyde fiberglass plastic in transverse compression.

where η_A pertains to axial creep in the matrix and $\eta_\perp$ refers to the transverse direction. The quantity $\eta_\perp$ is determined by the strain ratio at steady creep $\langle \varepsilon_\perp(\infty) \rangle$ and short-term loading $\langle \varepsilon_\perp(0) \rangle$ for a fixed value of the mean stress $\langle \sigma_1 \rangle$. Hence

$$\eta_\perp = \frac{E_\perp(\infty)}{E_\perp}. \tag{7.38}$$

The transverse modulus of elasticity $E_\perp(\infty)$ corresponds to the elastic modulus of the matrix at $t = \infty$, i.e.,

$$E_A(\infty) = E_A \frac{\beta_A}{\lambda_A + \beta_A}.$$

The strain is the nth layer involving both the fibers and the matrix can be used to obtain the axial stress

$$\sigma_{\perp n} = \frac{E_{Br}}{E_\perp} \left[1 - \frac{1 - \eta_\perp}{1 - \eta_A} \left(1 - \frac{E_A}{E_{Br}} \right) \right] \langle \sigma_\perp \rangle. \tag{7.39}$$

For the entire composite, there results

$$\langle \varepsilon_{\perp}(t) \rangle = \frac{1}{E_{\perp}} \cdot \frac{1}{1-\eta_A}[\eta_{\perp} - \eta_A + (1-\eta_{\perp})E_A D(t)]\langle \sigma_{\perp} \rangle. \tag{7.40}$$

The creep function of the matrix $D(t)$ is known from Equation (7.9). It has been shown that creep along the fiber direction is negligible in contrast to that in the transverse direction under longitudinal shear. The transverse strain can thus be assumed to be independent of the duration of load if the load is applied along the fibers. The reverse is also assumed to hold if the load is applied in the transverse direction. This is the result of symmetry of the compliance matrix for a unidirectional reinforced plastic.

In this way, the deformation of a unidirectional reinforced layer under sustained constant load and plane stress along the axes of elastic symmetry, can be described by

$$\begin{aligned} \langle \varepsilon_{11}(t) \rangle &= s_{11}\langle \sigma_{11} \rangle + \langle \sigma_{\perp} \rangle \\ \langle \varepsilon_{\perp}(t) \rangle &= s_{12}\langle \sigma_{11} \rangle + s_{22}(t)\langle \sigma_{\perp} \rangle \\ \langle \nu_{11\perp}(t) \rangle &= s_{66}(t)\langle \tau_{11\perp} \rangle. \end{aligned} \tag{7.41}$$

Here, s_{11} and s_{12} are the elastic constituents of a compliance matrix. The constituents $s_{22}(t)$ and $s_{66}(t)$ characterize the viscoelastic properties of unidirectional reinforced plastics and they are given by Equations (7.35) and (7.32).

The viscoelastic properties of a unidirectional reinforced plastic loaded in the fiber direction other than that of elastic symmetry, are of practical interest. Under plane stress, the time dependent strain and stress relations are given by

$$\begin{bmatrix} \langle \varepsilon_x(t) \rangle \\ \langle \varepsilon_y(t) \rangle \\ \langle \nu_{xy}(t) \rangle \end{bmatrix} = \begin{bmatrix} \overline{s}_{11}(t) & \overline{s}_{12}(t) & \overline{s}_{16}(t) \\ \overline{s}_{12}(t) & \overline{s}_{22}(t) & \overline{s}_{26}(t) \\ \overline{s}_{16}(t) & \overline{s}_{26}(t) & \overline{s}_{66}(t) \end{bmatrix} \begin{bmatrix} \langle \sigma_x \rangle \\ \langle \sigma_y \rangle \\ \langle \tau_{xy} \rangle \end{bmatrix}. \tag{7.42}$$

The viscoelastic compliances $\overline{s}_{ij}(t)$ in any direction can be obtained by the law of transformation of axes through the angle, say α. That is

$$\begin{aligned} \overline{s}_{11}(t) &= s_{11}\cos^4\alpha + [2s_{12} + s_{66}(t)]\sin^2\alpha\cos^2\alpha + s_{22}(t)\sin^4\alpha \\ \overline{s}_{12}(t) &= s_{12}(\sin^4\alpha + \cos^4\alpha) + [s_{11} + s_{22}(t) - s_{66}(t)]\sin^2\alpha\cos^2\alpha \\ \overline{s}_{22}(t) &= s_{11}\sin^4\alpha + [2s_{12} + s_{66}(t)]\sin^2\alpha\cos^2\alpha + s_{22}(t)\cos^4\alpha \\ \overline{s}_{16}(t) &= \left[s_{11} - s_{12} - \frac{1}{2}s_{66}(t)\right]\sin 2\alpha\cos^2\alpha + \left[s_{12} + \frac{1}{2}s_{66}(t) - \right. \\ &\quad \left. - s_{22}(t)\right]\sin 2\alpha\sin^2\alpha \\ \overline{s}_{26}(t) &= \left[s_{11} - s_{12} - \frac{1}{2}s_{66}(t)\right]\sin 2\alpha\sin^2\alpha + \left[s_{12} + \frac{1}{2}s_{66}(t) - \right. \end{aligned} \tag{7.43}$$

$$- s_{22}(t)\Big] \sin 2\alpha \cos^2\alpha$$

$$\overline{s}_{66}(t) = [s_{11} - 2s_{12} + s_{22}(t) - s_{66}(t)] \sin^2 2\alpha + s_{66}(t).$$

Note that if two of the s_{ij} in Equation (7.41) are time independent, then all $\overline{s}_{ij}$'s depend on time. These functions can be determined from the elastic properties of the constituents and $D(t)$ under axial loading. From Equations (7.32) and (7.35) with α being the direction of reinforcement, it can be shown that

$$\begin{aligned}
\overline{s}_{11}(t) &= a_{11} + b_{11}D(t) \quad & \overline{s}_{12}(t) &= a_{12} + b_{12}D(t) \\
\overline{s}_{22}(t) &= a_{22} + b_{22}D(t) \quad & \overline{s}_{16}(t) &= a_{16} + b_{16}D(t), \\
\overline{s}_{26}(t) &= a_{26} + b_{26}D(t) \quad & \overline{s}_{66}(t) &= a_{66} + b_{66}D(t)
\end{aligned} \tag{7.44}$$

where

$$\begin{aligned}
a_{11} &= s_{11} \cos^4\alpha + d_1 \sin^4\alpha + (2s_{12} + g_1) \sin^2\alpha \cos^2\alpha \\
a_{12} &= s_{12}(\sin^4\alpha + \cos^4\alpha) + (s_{11} + d_1 - g_1) \sin^2\alpha \cos^2\alpha \\
a_{22} &= s_{11} \sin^4\alpha + d_1 \cos^4\alpha + (2s_{12} + g_1) \sin^2\alpha \cos^2\alpha \\
a_{16} &= \left[\left(s_{11} - s_{12} - \frac{1}{2}g_1\right) \cos^2\alpha + \left(s_{12} + \frac{1}{2}g_1 - d_1\right) \sin^2\alpha\right] \sin 2\alpha \\
a_{26} &= \left[\left(s_{11} - s_{12} - \frac{1}{2}g_1\right) \sin^2\alpha + \left(s_{12} + \frac{1}{2}g_1 - d_1\right) \cos^2\alpha\right] \sin 2\alpha \\
a_{66} &= (s_{11} - 2s_{12} + d_1 - g_1) \sin^2 2\alpha + g_1 \\
b_{11} &= [d_2 \sin^2\alpha + g_2 \cos^2\alpha] \sin^2\alpha \\
b_{12} &= (d_2 - g_2) \sin^2\alpha \cos^2\alpha \\
b_{22} &= [d_2 \cos^2\alpha + g_2 \sin^2\alpha] \cos^2\alpha \\
b_{26} &= [(g_2 - d_2) \cos^2\alpha - g_2 \sin^2\alpha] \sin 2\alpha \\
b_{16} &= [(g_2 - d_2) \sin^2\alpha - g_2 \cos^2\alpha] \sin 2\alpha \\
b_{66} &= (d_2 - g_2) \sin 2\alpha + g_2.
\end{aligned}$$

7.4 Creep of bidirectional reinforced plastics

Consider plastics that are made of unidirectional reinforced layers of equal thickness with the directions of alternating fibers by the angle 2α in two successive layers. Such a material is orthotropic with the axes of elastic symmetry directed at angles α and $90^\circ - \alpha$ to the directions of reinforcement. Bidirectional reinforcement can be made from symmetrically oriented helical winding on a cylindrical mandrel.

A particular case of bidirectional reinforced materials is an orthogonally reinforced plastic with an equal volume fraction of layers, oriented in two mutually perpendicular directions. The axes of elastic symmetry are oriented at an angle of 45° to the directions of reinforcement.

Under plane stress, the strain field of a bidirectional reinforced plastic under sustained constant load along the axes of elastic symmetry is given by

$$\begin{bmatrix} \langle\langle \varepsilon_x(t) \rangle\rangle \\ \langle\langle \varepsilon_y(t) \rangle\rangle \\ \langle\langle \nu_{xy}(t) \rangle\rangle \end{bmatrix} = \begin{bmatrix} \overset{*}{s}_{11}(t) & \overset{*}{s}_{12}(t) & 0 \\ \overset{*}{s}_{12}(t) & \overset{*}{s}_{22}(t) & 0 \\ 0 & 0 & \overset{*}{s}_{66}(t) \end{bmatrix} \begin{bmatrix} \langle\langle \sigma_x \rangle\rangle \\ \langle\langle \sigma_y \rangle\rangle \\ \langle\langle \tau_{xy} \rangle\rangle \end{bmatrix}. \tag{7.45}$$

Refer to the example in Figure 7.9. The quantities $\overset{*}{s}_{ij}(t)$ are related to the properties of the unidirectional reinforced layers, oriented at angles $\pm\alpha$ to the direction of $\langle\langle \sigma_x \rangle\rangle$. The determination of $\overset{*}{s}_{ij}(t)$ involves difficulties associated with the system of integral equations governing the strains and stresses in the unidirectional reinforced layers. Even under a constant external load, the stresses in the bidirectional reinforced plastics vary with time. Thus, regardless of the direction of loading, all stresses are time dependent and determined from the condition that $\langle \nu_{xy} \rangle = \langle\langle \sigma_{xy} \rangle\rangle = 0$. The maximum value of stresses $\langle \tau_{xy} \rangle$ are determined from

$$\langle \tau_{xy} \rangle = \frac{-\overline{s}_{16}}{\overline{s}_{66}} \langle\langle \sigma_x \rangle\rangle \tag{7.46}$$

where $\overline{s}_{16}$ and $\overline{s}_{66}$ are related to the elastic properties of a unidirectional reinforced plastic.

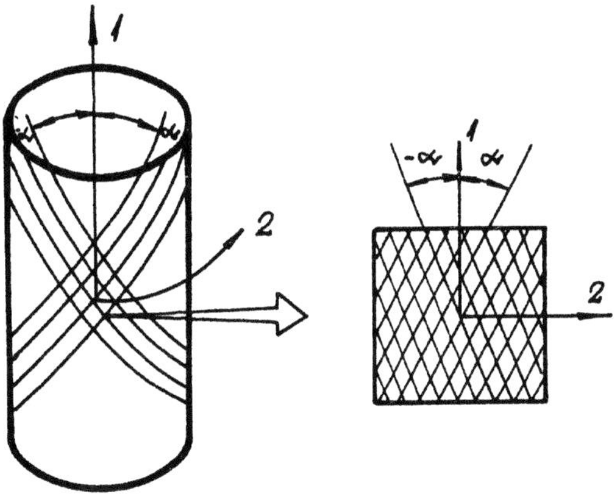

Fig. 7.9: Bidirectional angle-ply reinforced plastic.

If $\overline{s}_{16}$ and $\overline{s}_{66}$ in Equation (7.46) are substituted respectively by $\overline{s}_{16}(\infty)$ and $\overline{s}_{66}(\infty)$, then Equation (7.44) gives the maximum stress $\langle \tau_{xy} \rangle_\infty$. Evaluation of the viscoelastic compliance matrix is simplified by assuming that $\langle \tau_{xy} \rangle$ remains constant. This leads to

$$s_{11}^{*}(t) = \bar{s}_{11}(t) - \frac{\bar{s}_{16}^{2}(t)}{\bar{s}_{66}(t)}$$

$$s_{12}^{*}(t) = \bar{s}_{12}(t) - \frac{\bar{s}_{16}(t)\bar{s}_{26}(t)}{\bar{s}_{66}(t)} \qquad (7.47)$$

$$s_{22}^{*}(t) = \bar{s}_{22}(t) - \frac{\bar{s}_{26}^{2}(t)}{\bar{s}_{66}(t)}.$$

The creep deformation of a bidirectionally reinforced plastic under constant mean stress $\langle\langle \sigma_x \rangle\rangle$ becomes

$$\langle\langle \varepsilon_x(t) \rangle\rangle = \left[\bar{s}_{11}(t) - \frac{\bar{s}_{16}^{2}(t)}{\bar{s}_{66}(t)} \right] \langle\langle \sigma_x \rangle\rangle. \qquad (7.48)$$

If the functions $\bar{s}_{11}(t)$, $\bar{s}_{16}(t)$ and $\bar{s}_{66}(t)$ are described by Equation (7.44) and expressed in terms of the function $D(t)$, the creep deformation of a bidirectionally reinforced plastic may be expressed by one experimentally obtained creep function for a matrix under axial loading. In a particular case, if $\alpha = 45°$, the creep of a cross-ply laminate at an angle 45° to the direction of reinforcement can be found from Equation (7.48):

$$\bar{s}_{11}(t) = \frac{1}{4}(s_{11} + 2s_{12} + d_1 + g_1) + \frac{1}{4}(d_2 + g_2)D(t)$$

$$\bar{s}_{16}(t) = \frac{1}{2}(s_{11} - d_1) - \frac{1}{2}D(t) \qquad (7.49)$$

$$\bar{s}_{66}(t) = s_{11} - 2s_{12} + d_1 + d_2 D(t).$$

Figure 7.10 displays the experimental data and calculated creep curves for a cross-ply phenolformaldehyde-fiberglass laminate under compression at an angle 45° to the direction of reinforcement. The experimental results are based on a compressive stress of $\langle\langle \sigma_x \rangle\rangle = 57$ MPa.

7.5 Stress-rupture under sustained load

Consider an M-directional plastic having a symmetrical structure relative to the midplane under an applied constant load. The state of stress in the layers at time t is given by [38]:

$$\langle \sigma_{11}(t) \rangle_i = \langle \sigma_{11}(O) \rangle_i f_{11i}(t)$$

$$\langle \sigma_{\perp}(t) \rangle_i = \langle \sigma_{\perp}(O) \rangle_i f_{\perp i}(t) \qquad (7.50)$$

$$\langle \tau_{11\perp}(t) \rangle_i = \langle \tau_{11\perp}(O) \rangle_i f_{11\perp i}(t)$$

where $i = 1, \ldots, M$.

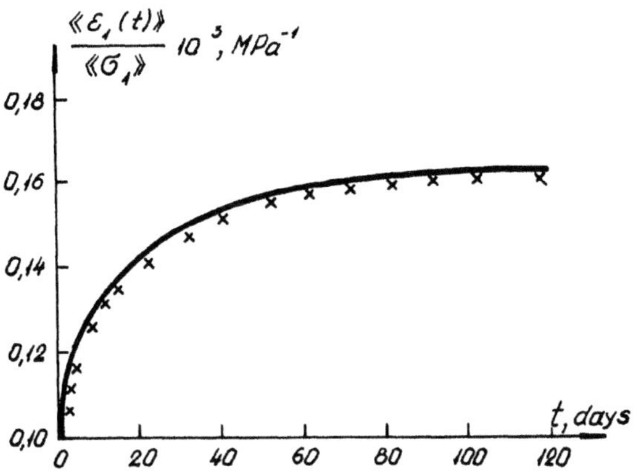

Fig. 7.10: Creep of equal strength orthogonally reinforced phenolformaldehyde fiberglass plastic in uniaxial compression at an angle 45° to the direction of reinforcement.

The long-term strength behavior of reinforced plastics can be obtained from Equation (7.50) by a set of discrete values of $\langle \sigma_j(t) \rangle_i$, where $j = 11, \perp, 11\perp$ and $i = 1, ..., M$, for a definite interval Δt. The surface of this laminate will be determined by the interfaces of the layers within which $\langle\langle \sigma_1 \rangle\rangle$, $\langle\langle \sigma_2 \rangle\rangle$ and $\langle\langle \tau_{12} \rangle\rangle$ prevail, where $\langle\langle \sigma_1 \rangle\rangle$, $\langle\langle \sigma_2 \rangle\rangle$ and $\langle\langle \tau_{12} \rangle\rangle$ are the applied loads. The fracture of fiber or matrix may cause a layer to fail; the stress rupture and sustained load of a layer that has failed by debonding has been discussed [39]. This occurs when the bend strength exceeds the matrix strength. The long-term strength of a laminate is defined in terms of loss of continuity.

Failure of the entire plastic is caused by fiber fracture in the ith layer ($i = 1, ..., M$).

It was shown in [25, 28, 40] that damage accumulation $\partial w / \partial t$ is a function of many parameters. For linear accumulation of damage, $\partial w / \partial t$ is independent of the damage level. By neglecting the creep of a layer in the direction of reinforcement, the following strength criterion [41] is obtained:

$$\int_0^{t_*} \frac{d\tau}{t_{11}^{+(-)}[\langle \sigma_{11}(\tau) \rangle_i]} = 1 \tag{7.51}$$

Assuming that all the load is taken by the fibers, Equation (7.51) becomes

$$\psi \int_0^{t_*} \frac{d\tau}{t_B^{+(-)}[\langle \sigma_{11}(\tau) \rangle_i]} = 1 \tag{7.52}$$

where $t_{11}^{+(-1)}(\sigma)$ and $t_B^{+(-1)}(\sigma)$ are time to failure in tension (compression) of a layer and fibers, respectively, with ψ being the volume fraction. For convenience of integration, Equation (7.52) can be expressed as

$$\frac{1}{t_B^{+(-)}(\sigma)} = \sum_{i=1}^{N} b_i \left[\frac{\sigma - R_B^{+(-)}(\infty)}{R_B^{+(-)}(O) - \sigma}\right]^i;$$
$$R_B^{+(-)}(\infty) \le \sigma \le R_B^{+(-)}(O) \qquad (7.53)$$

where $R_B^{+(-)}(O)$ and $R_B^{+(-)}(\infty)$ are the short-term and long-term ultimate strength of fibers in uniaxial tension (compression), respectively, where b_i $(i = 1, ..., N)$ are material constants. The values $R_B^{+(-)}(O)$ and $R_B^{+(-)}(\infty)$, b_i $(i = 1, ..., N)$ are determined by experiments under constant load. It can be concluded that

Failure of the entire plastic is caused by the matrix in the ith layer $(i = 1, ..., N)$.

The energy criterion in [42] gives

$$F^*\{\langle\sigma_\perp(t)\rangle_i;\ \langle\tau_{11\perp}(t)\rangle_i\} = 1. \qquad (7.54)$$

A more elaborate version would be [43, 44] at $|\langle\tau_{11\perp}(t)\rangle_i| / \langle\sigma_\perp(t)\rangle_i \le a_1^*(t)$; (zone AB, Fig. 7.11)

$$\psi^*\{\langle\sigma_\perp(t)\rangle_i^2 \cdot a_2^*(t) \mid \langle\tau_{11\perp}(t)\rangle_i^2 \cdot a_3^*(t)\} = (R_A^+)^2 \qquad (7.55)$$

at $|\langle\tau_{11\perp}(t)\rangle_i| / \langle\sigma_\perp(t)\rangle_i \ge a_1^*(t)$ (zone BC, Fig. 7.11)

$$\psi^*\{\langle\sigma_\perp(t)\rangle_i^2 \cdot a_4^*(t) + \langle\sigma(t)\rangle_i\langle\tau_{11\perp}(t)\rangle_i \cdot a_5^*(t) + $$
$$+ \langle\tau_{11\perp}(t)\rangle_i^2 \cdot a_6^*(t)\} = (R_A^+)^2 \qquad (7.56)$$

at $0 \ge |\langle\tau_{11\perp}(t)\rangle_i| / \langle\sigma_\perp(t)\rangle_i \ge -a_1^*(t)$ (zone DE, Fig. 7.11)

$$\psi^*\{\langle\sigma_\perp(t)\rangle_i^2 \cdot a_2^*(t) + \langle\tau_{11\perp}(t)\rangle_i^2 \cdot a_3^*(t)\} = (R_A^-)^2 \qquad (7.57)$$

at $|\langle\tau_{11\perp}(t)\rangle_i| / \langle\sigma_\perp(t)\rangle_i \le -a_1^*(t)$ (zone EFC, Fig. 7.11)

$$\psi^*\{\langle\sigma_\perp(t)\rangle_i^2 a_4^*(t) - \langle\sigma_\perp(t)\rangle_i\langle\tau_{11\perp}(t)\rangle_i\, a_5^*(t) +$$
$$\langle\tau_{11\perp}(t)\rangle_i^2 a_6^*(t)\} = (R_A^-)^2 \qquad (7.58)$$
$$\psi^*\{\langle\sigma_\perp(t)\rangle_i^2 \cdot a_7^*(t) + \langle\sigma_\perp(t)\rangle_i\langle\tau_{11\perp}(t)\rangle_i\, a_5^*(t) +$$
$$\langle\tau_{11\perp}(t)\rangle_i^2 a_6^*(t)\} = (R_A^+)^2$$

where ψ^* is the operator:

$$\psi^*[F(t)] = F(t) + 2\int_0^t k_A(t-\tau)F(\tau)\,d\tau. \qquad (7.59)$$

Here, $R_A^{+(-)}$ and $k_A(t-\tau)$ are the short-term ultimate strength in tension (compression) and the kernel of the matrix, respectively. The expressions $a_k^*(t)$, $k = 1, ..., 7$ are determined from [44]

$$a_1^*(t) = \sqrt{\nu_A\overline{\sigma}_A(t)\cdot[\overline{\sigma}_{A\perp}(t) + \sigma_{A\parallel}(t)]}$$
$$a_2^*(t) = \{\overline{\sigma}_{A\perp}^2(t) + \overline{\sigma}_{A\parallel}^2(t)\}(1-\nu_A^2) - 2\nu_A(1+\nu_A)\overline{\sigma}_{A\perp}(t)\overline{\sigma}_{A\parallel}(t)$$

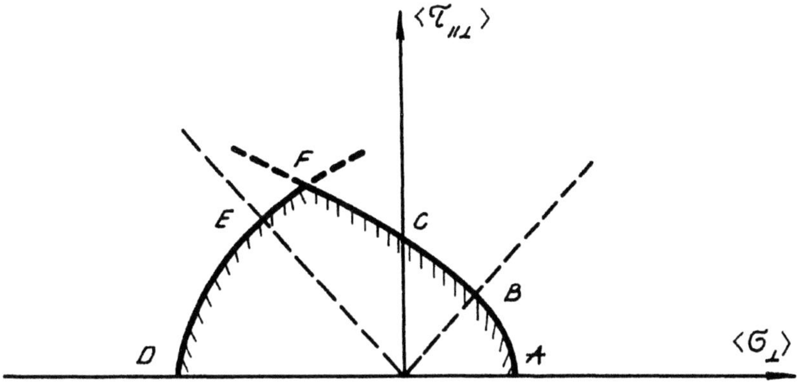

Fig. 7.11: Strength envelope of a unidirectional reinforced layer under combined loading.

$$a_3^*(t) = 2(1+\nu_A) \tag{7.60}$$
$$a_4^*(t) = \{\overline{\sigma}^2_{A\perp}(t) + 2\overline{\sigma}^2_{A\parallel}(t)\}(1-\nu_A)^2\cdot 0.5 - 1.5\nu_A(1+\nu_A)\overline{\sigma}_{A\perp}(t)\overline{\sigma}_{A\parallel}(t)$$
$$a_5^*(t) = (1+\nu_A)\overline{\sigma}_{A\perp}(t)$$
$$a_6^*(t) = (1+\nu_A) = 0.5a_3^*(t)$$
$$a_7^*(t) = \overline{\sigma}^2_{A\perp}(t)(1-\nu_A^2)\cdot 0.5 - \overline{\sigma}_{A\perp}(t)\overline{\sigma}_{A\parallel}(t)\nu_A(1+\nu_A)\cdot 0.5$$

where ν_A is the Poisson ratio of the matrix, $\overline{\sigma}_{A\perp}(t)$ and $\overline{\sigma}_{A\parallel}(t)$ are related to the viscoelastic properties of the constituents and their distribution [35, 37].

Consider the long-term ultimate state of an angle-ply reinforced plastic under uniaxial loading with a constant load in the '1' direction of the axis of elastic symmetry such that the stresses $\langle\langle\sigma_2\rangle\rangle = \langle\langle\tau_{12}\rangle\rangle = 0$. The stress state in the layers of the plastic is

$$\begin{aligned}\langle\sigma_{11}(t)\rangle &= \langle\langle\sigma_1\rangle\rangle[\cos^2\alpha - 2\sin\alpha\cos\alpha\cdot v_{12}(t)]\\ \langle\sigma_{\perp}(t)\rangle &= \langle\langle\sigma_1\rangle\rangle[\sin^2\alpha + 2\sin\alpha\cos\alpha\cdot v_{12}(t)]\\ \langle\tau_{11\perp}(t)\rangle &= \langle\langle\sigma_1\rangle\rangle[\sin\alpha\cos\alpha + (\cos^2\alpha - \sin^2\alpha)v_{12}(t)]\end{aligned} \tag{7.61}$$

where

$$v_{12}(t) = \frac{\langle\tau_{12}(t)\rangle}{\langle\langle\sigma_1\rangle\rangle}. \tag{7.62}$$

Refer to Figure 7.12. Failure corresponds to $R_1 = \langle\langle\sigma_1\rangle\rangle$. Equation (7.52) gives the strength of an angle-ply reinforced plastic that corresponds to fiber fracture:

$$\psi\int_0^{t_*}\frac{d\tau}{t_B^{+(-)}\{R_1^+[\cos^2\alpha - 2\sin\alpha\cos\alpha\cdot v_{12}(\tau)\}} = 1. \tag{7.63}$$

For $t = \infty$, it gives

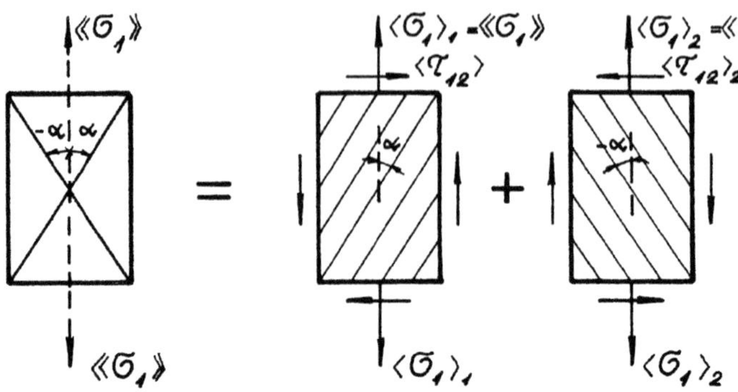

Fig. 7.12: Stress state in layers of an angle-ply reinforced plastic under uniaxial loading.

$$R_1(0,\infty) = \frac{\psi R_B^{+(-)}(0,\infty)}{\cos^2\alpha - 2\sin\alpha\cos\alpha\,\nu_{12}(0,\infty)}. \tag{7.64}$$

Fracture of the matrix occurs at $\langle \sigma_\perp(t) \rangle > 0$, this corresponds to $\langle\langle\sigma_1\rangle\rangle \nu_{12}(t) > -0.5\tan\alpha \langle\langle\sigma_1\rangle\rangle$ and Equations (7.55) and (7.56) yield

(1) at

$$\nu_{12}(t) \le \frac{a_1^*(t)\sin^2\alpha - \sin\alpha\cos\alpha}{\cos^2\alpha - \sin^2\alpha - 2a_1^*(t)\sin\alpha\cos\alpha}$$

(zone AB, Fig. 7.11)

$$R_1(t) = \frac{R_A^+}{\psi^*\{G_1 + G_2\nu_{12}(t) + G_3\nu_{12}^2(t)\}^{1/2}} \tag{7.65}$$

where

$$G_1 = a_2^*(t)\sin^4\alpha + a_3^*(t)\sin^2\alpha\cos^2\alpha$$

$$G_2 = a_2^*(t)4\sin^3\alpha\cos\alpha + a_3^*(t)2\sin\alpha\cos\alpha(\cos^2\alpha - \sin^2\alpha)$$

$$G_3 = a_2^*(t)4\sin^2\alpha\cos^2\alpha + a_3^*(t)(\cos^2\alpha - \sin^2\alpha)$$

(2) at

$$\nu_{12}(t) > \frac{a_1^*(t)\sin^2\alpha - \sin\alpha\cos\alpha}{\cos^2\alpha - \sin^2\alpha - 2a_1^*(t)\sin\alpha\cos\alpha}$$

(zone BC, Fig. 7.11)

$$R_1(t) = \frac{R_A^+}{\psi^*\{G_4 + G_5\nu_{12}(t) + G_6\nu_{12}^2(t)\}^{1/2}} \tag{7.66}$$

where

$$G_4 = a_4^*(t)\sin^4\alpha + a_5^*(t)\sin^3\alpha\cos\alpha + a_6^*(t)\sin^2\alpha\cos^2\alpha$$

$$G_5 = a_4^*(t)4\sin^3\alpha\cos\alpha + a_5^*(t)(3\sin^2\alpha\cos^2\alpha - \sin^4\alpha) + \\ + a_6^*(t)2\sin\alpha\cos\alpha\cdot(\cos^2\alpha - \sin^2\alpha)$$

$$G_6 = a_4^*(t)4\sin^2\alpha\cos^2\alpha + a_5^*(t)2\sin\alpha\cos\alpha(\cos^2\alpha - \sin^2\alpha) + \\ + a_6^*(t)(\cos^2\alpha - \sin^2\alpha)^2.$$

For $\langle\sigma_\perp(t)\rangle < 0$ and $\nu_{12}(t)\langle\langle\sigma_1\rangle\rangle < -0.5\tan\alpha\langle\langle\sigma_1\rangle\rangle$, Equations (7.57) and (7.58) render

(1) at

$$\nu_{12}(t) \geq \frac{a_1^*(t)\sin^2\alpha - \sin\alpha\cos\alpha}{\cos^2\alpha - \sin^2\alpha - 2a_1^*(t)\sin\alpha\cos\alpha}$$

(zone DE, Fig. 7.11)

$$R_1(t) = \frac{R_A^-}{\psi^*\{G_1 + G_2\nu_{12}(t) + G_3\nu_{12}^2(t)\}^{1/2}} \tag{7.67}$$

(2) at

$$\nu_{12}(t) < \frac{a_1^*(t)\sin^2\alpha - \sin\alpha\cos\alpha}{\cos^2\alpha - \sin^2\alpha - 2a_1^*(t)\sin\alpha\cos\alpha}$$

$$R_1(t) = \min\{R_1^{(1)}(t); R_1^{(2)}(t)\} \tag{7.68}$$

where

$$R_1^{(1)}(t) = \frac{R_A^-}{\psi^*\{G_7 + G_8\nu_{12}(t) + G_9\nu_{12}^2(t)\}^{1/2}} \tag{7.69}$$

$$R_1^{(2)}(t) = \frac{R_A^+}{\psi^*\{G_{10} + G_{11}\nu_{12}(t) + G_{12}\nu_{12}^2(t)\}^{1/2}} \tag{7.70}$$

Here,

$$G_7 = a_4^*(t)\sin^4\alpha - a_5^*(t)\sin^3\alpha\cos\alpha + a_6^*(t)\sin^2\alpha\cos^2\alpha$$

$$G_8 = a_4^*(t)4\sin^3\alpha\cos\alpha - a_5^*(t)(3\sin^2\alpha\cos^2\alpha - \sin^4\alpha) + \\ + a_6^*(t)2\sin\alpha\cos\alpha(\cos^2\alpha - \sin^2\alpha)$$

$$G_9 = a_4^*(t)4\sin^2\alpha\cos^2\alpha - a_5^*(t)2\sin\alpha\cos\alpha(\cos^2\alpha - \sin^2\alpha) + \\ + a_6^*(t)(\cos^2\alpha - \sin^2\alpha)^2$$

$$G_{10} = a_7^*(t)\sin^4\alpha + a_5^*(t)\sin^3\alpha\cos\alpha + a_6^*(t)\sin^2\alpha\cos^2\alpha$$

$$G_{11} = a_7^*(t)4\sin^3\alpha\cos\alpha + a_5^*(t)(3\sin^2\alpha\cos^2\alpha - \sin^4\alpha) + \\ + a_6^*(t)2\sin\alpha\cos\alpha(\cos^2\alpha - \sin^2\alpha)$$

$$G_{12} = a_7^*(t)4\sin^2\alpha\cos^2\alpha + a_5^*(t)2\sin\alpha\cos\alpha(\cos^2\alpha - \sin^2\alpha) +$$

$$+ a_6^*(t)(\cos^2\alpha - \sin^2\alpha)^2.$$

The ultimate values of R_1 at $t_1 \to 0$ and $t \to \infty$ can be obtained as

$$\begin{aligned}
&a_i^*(t)_{i=1,\ldots,7} \to a_i(0, \infty)\\
&\nu_{12}(t) \to \nu_{12}(0, \infty)\\
&\psi^*[F(t)] \to F(0), F(\infty)\left[1 + 2\int_0^{\infty} k(t-\tau)\,d\tau\right]
\end{aligned} \tag{7.71}$$

which can be substituted into Equations (7.65) and (7.70). Figure 7.13 compares the experimental data [45] of the long-term strength in tension of an angle-ply reinforced boron plastic at $\alpha = 45^\circ$ with the theoretical predictions obtained from Equations (7.65) and (7.70) for $E_B/E_A = 100$, $\psi = 0.60$, $\nu_A = 0.35$, $\nu_B = 0.23$, $k_A(t-\tau) = c_A e^{-d_A(t-\tau)}$, $c_A = 0.0033\,(\text{min})^{-1}$, and $d_A = 0.0289\,(\text{min})^{-1}$. Note that fiber fracture occurs only at small angles α and, as a rule, the failure of such laminates is caused by failure of the matrix or by debonding. Under uniaxial loading ($\langle\langle\tau_{12}\rangle\rangle = 0$) the stress state in the layers are

$$\begin{aligned}
\langle\sigma_{11}(t)\rangle &= \langle\langle\sigma_1\rangle\rangle[\cos^2\alpha - 2\sin\alpha\cos\alpha\,\nu_{12}(t)] +\\
&\quad + \langle\langle\sigma_2\rangle\rangle[\sin^2\alpha - 2\sin\alpha\cos\alpha\,\nu_{12}^1(t)]\\
\langle\sigma_{\perp}(t)\rangle &= \langle\langle\sigma_1\rangle\rangle[\sin^2\alpha + 2\sin\alpha\cos\alpha\,\nu_{12}(t)] +\\
&\quad + \langle\langle\sigma_2\rangle\rangle[\cos^2\alpha + 2\sin\alpha\cos\alpha\,\nu_{12}^1(t)]\\
\langle\tau_{11\perp}(t)\rangle &= \langle\langle\sigma_1\rangle\rangle[\sin\alpha\cos\alpha + (\cos^2\alpha - \sin^2\alpha)\nu_{12}(t)] +\\
&\quad + \langle\langle\sigma_2\rangle\rangle[\sin\alpha\cos\alpha - (\cos^2\alpha - \sin^2\alpha)\nu_{12}^1(t)]
\end{aligned} \tag{7.72}$$

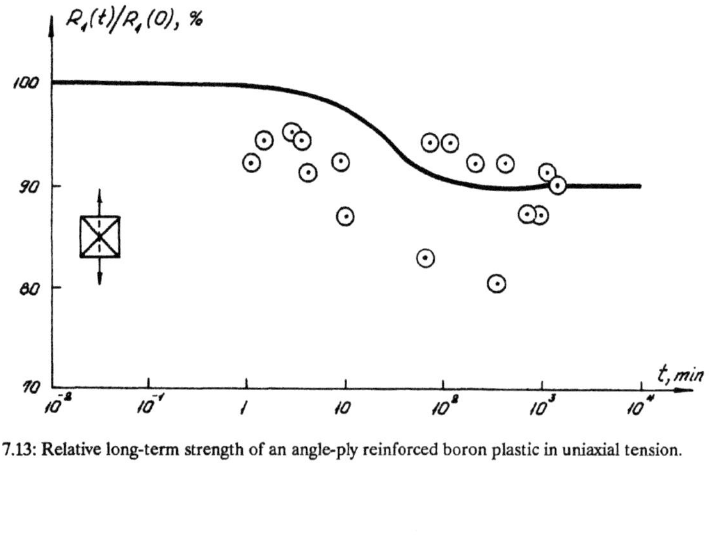

Fig. 7.13: Relative long-term strength of an angle-ply reinforced boron plastic in uniaxial tension.

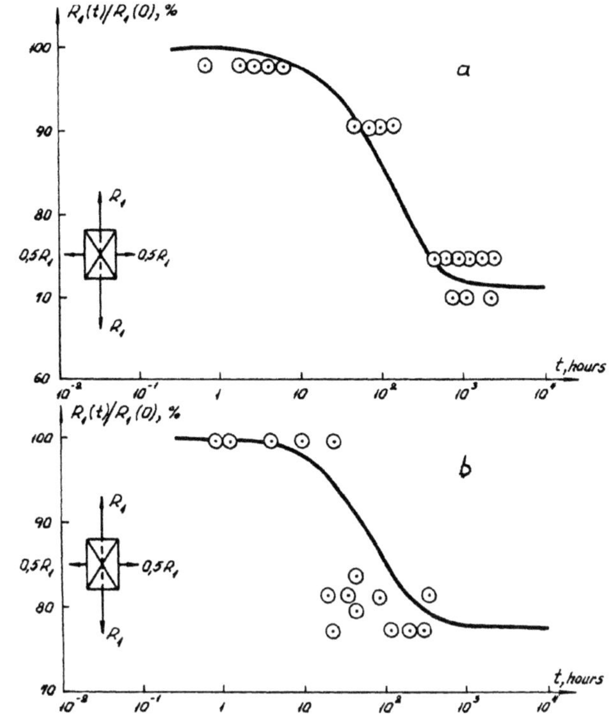

Fig. 7.14: Relative long-term strength of an angle-ply reinforced fiberglass plastic in biaxial tension: (a) epoxy and (b) polyester.

where $\nu^1_{12}(\alpha, t) = \nu_{12}(90^\circ - \alpha, t)$. Substitution of Equation (7.72) into Equations (7.51) and (7.58) give the strength envelope under biaxial stress. Figure 7.14 compares the experimental data [46] with the long-term strength of a helically wound tube at $\alpha = \pm 50^\circ$ made of fiberglass plastic where $E_B/E_A = 20$, $\psi = 0.60$, $c_A = 0.00245\ \text{hours}^{-1}$, $d_A = 0.00418\ \text{hours}^{-1}$ as given in Figure 7.14(a), and phenolformaldehyde-fiberglass plastic where $E_B/E_A = 20$, $\psi = 0.60$, $c_A = 0.00293\ \text{hours}^{-1}$, $d_A = 0.00763\ \text{hours}^{-1}$, as given in Figure 7.14(b). The internal pressures are $\langle\langle \sigma_1 \rangle\rangle = P$ and $\langle\langle \sigma_2 \rangle\rangle = 0.5P$.

Figure 7.13 shows that the relative variation in the strength of angle-ply laminated plastics under sustained loading is weaker than that of a unidirectional material of the same initial properties in each layer. This is due to redistribution of stresses in the layer which undergo viscoelastic deformation. As noted in [44], for constant load, the increase in stresses in the direction of reinforcement $\langle \sigma_{11}(t) \rangle_i$ is accompanied by a decrease in the stresses $\langle \sigma_\perp(t) \rangle_i$ and $\langle \tau_{11\perp}(t) \rangle_i$. This is attributed to the increase of the components of the compliance tensor of the layers

$$s^{(i)}_{22}(t) = E^{(i)}_{22}(t)^{-1} \quad \text{and} \quad s^{(i)}_{66}(t) = G^{(i)}_{11\perp}(t)^{-1}.$$

The strength of a laminated plastic decreases with time because of changes in material properties of the constituents; this, however, is to some extent compensated by the

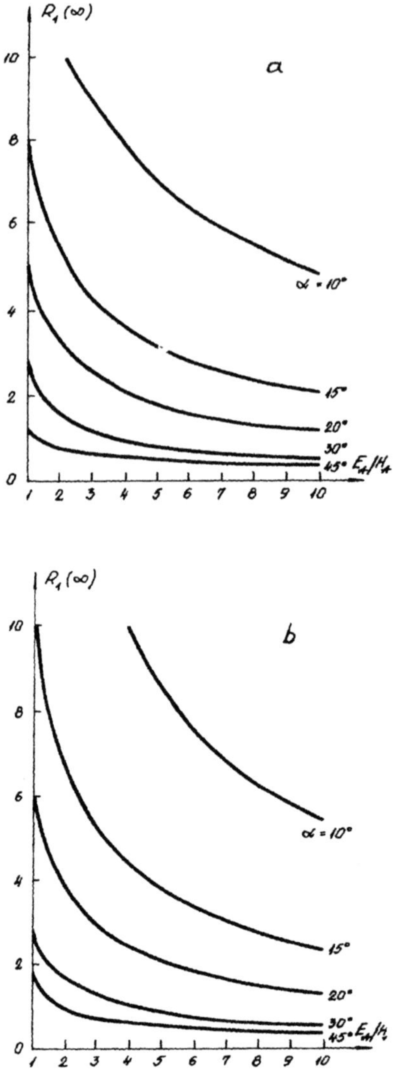

Fig. 7.15: Dependence of ultimate strength of an angle-ply reinforced plastics of the creep of a polymeric matrix in uniaxial tension: (a) fiberglass and (b) carbon.

laminate structure geometry. This can be seen from the tensile strength dependence on the creep of matrix for an angle-ply fiberglass with $E_B/E_A = 20$, and $\psi = 0.60$, carbon plastic with $E_{Br}/E_A = 2$, and $E_{Bz}/E_{Br} = 20$, $\psi = 0.60$, and $R(t \to \infty)$. Failure of the entire plastic is caused by the fracture of the matrix. Figure 7.15 shows that if the E_A/H_A increases beyond all bounds, where H_A is the long-term modulus of elasticity of the matrix, then the values of $R_1(t \to \infty)$ of an angle-ply reinforced plastic tend to some asymptotes, other than zero. This condition becomes particularly important if the ultimate strength of a layer tends to zero at $E_A/H_A \to \infty$.

Note that the change of the surface of long-term strength of a laminated plastic is close to affine transformation. The anistropy of the long-term relative strength, $R_{ij}(t)/R_{ij}(0)$ becomes higher; rheological properties of the constituents then becomes more important.

The above approach for defining the long-term strength of reinforced plastics suggest a design procedure. Aside from selecting the appropriate strength according to service conditions, a minimum thickness of the laminated plastic for a given set of properties has been suggested [47]. Load duration can be another factor [48] that should be considered in terms of the strength of the constituents and that of the composite structure.

References

[1] MILEIKO, S.T. and RABOTNOV, Yu.N., *Advances in Mechanics,* Vol. 3, 1980, p. 3 (in Russian).

[2] CHEREVATSKII, S.B. and ROMASHOV, Yu.P., Study of Revolution Formed by One Family of Filaments, in: *Strength and Dynamics of Aero Engines,* Issue 4, Maschinostroyeniye Press, Moscow, 1966, p. 5 (in Russian).

[3] OBRAZTSOV, I.F., VASIL'EV, V.V. and BUNAKOV, V.A., *Optimal Reinforcement of Shells from Composite Materials,* Machinostroyeniye Press,Moscow, 1977 (in Russian).

[4] PROTASOV, V.D., *Journal of the Mendeleyev All-Union Chemical Society,* Vol. 13, 1978, p. 289 (in Russian).

[5] VASIL'EV, V.V., *Mechanics of Composite Structures,* Mashinostroyeniye, Moscow, 1988 (in Russian).

[6] RABOTNOV, Y.N., *Herald of the USSR Academy of Sciences,* Vol. 33, 1965 (in Russian).

[7] RABOTNOV, Y.N., *Herald of the USSR Academy of Sciences,* Vol. 50, 1979 (in Russian).

[8] BOLOTIN, V.V. and NOVICHKOV, Yu.N., *Mechanics of Multilayer Structures,* Mashinostroyeniye Press, Moscow, 1980 (in Russian).

[9] CHRISTENSEN, R.M., *Mechanics of Composite Materials,* John Wiley and Sons, New York, 1979.

[10] VAN FO FI, G.A., *Reinforced Plastic Structures,* Tekhnika Press, Kiev, 1971 (in Russian).

[11] FEODOSYEV, V.I., *Ten Lectures - Talks on Strength of Materials,* Nauka Press, Moscow, 1969 (in Russian).

[12] LEKHNITSKII, S.G., *Anisotropic Plates,* Gordon and Breach, New York, 1968.

[13] LEKHNITSKII, S.G., *Theory of Elasticity of an Anisotropic Body,* 2nd ed., Nauka Press, Moscow, 1977 (in Russian).

[14] AMBARTSUMYAN, S.A., *Theory of Anisotropic Plates,* Technomic, Westport, Connecticut, 1970.

[15] AMBARTSUMYAN, S.A., *Proceedings of the Academy of Sciences of the Armenian SSR,* Vol. 21, 1968, p. 3 (in Russian).

[16] CHAMIS, C.C. (Ed.), *Structural Design and Analysis, Part I, Composite Materials,* ed. by L.J.

Broutman and R.H. Krock, Vol. 7, Academic Press, New York and London, 1974.

[17] CHAMIS, C.C. (Ed.), *Structural Design and Analysis, Part II, Composite Materials*, ed. by L.J. Broutman and R.H. Krock, Vol. 8, Academic Press, New York and London, 1974.

[18] TARNOPOL'SKII, Yu.M. and ROZE, A.V., *Characteristics of Calculation of Reinforced Plastic Parts*, Zinatne Press, Riga, 1969 (in Russian).

[19] TARNOPOL'SKII, Yu.M., ZHIGUN, I.G. and POLYAKOV, V.A., *Spatially Reinforced Composite Materials*, 2nd Edition: Mashinostroiyeniye, Moscow, 1987 (in Russian).

[20] RABOTNOV, Yu.N., *Elements of Hereditary Mechanics of Solids*, Nauka Press, Moscow, 1977 (in Russian).

[21] TARNOPOL'SKII, Yu.M., PORTNOV, G.G. and BEIL, A.I., Technological Problems in Composite Winding, in: *Mechanics of Composites*, Mir Publishers, Moscow, 1982, p. 186.

[22] TARNOPOL'SKII, Yu.M. and SKUDRA, A.M., *Construction Strength and Deformation Properties of Fiber Glass Reinforced Plastics*, U.S. Army Engineer Research and Development Lab., 1966.

[23] TARNOPOL'SKII, Yu.M. and BEIL, A.I., Problems of the Mechanics of Composite Winding, in: *Fabrication of Composites*, ed. by A. Kelly and S.T. Mileiko, Handbook of Composites, Vol. IY, 56, North Holland Publishing Co., Amsterdam and New York.

[24] RABOTNOV, Yu.N., *Applied Mechanics and Mathematics*, Vol. 12, 1948, p. 53 (in Russian).

[25] RABOTNOV, Yu.N., *Creep of Structural Elements*, Nauka Press, Moscow, 1966 (in Russian).

[26] RAZHANITSIN, A.R., *Theory of Creep*, Stroyizdat Press, Moscow, 1969 (in Russian).

[27] RIVKIND, V.N., Candidate's Dissertation, Leningrad Institute of Shipbuilding, Leningrad, 1969 (in Russian).

[28] RABOTNOV, Yu.N., *Creep of Structural Elements*, Nauka Press, Moscow, 1966 (in Russian).

[29] RABOTNOV, Yu.N., PAPERNIK, L.H. and ZVONOV, E.N., *Tables of Fractionally-Exponential Function for Negative Parameter and Its Integral*, Nauka Press, Moscow, 1969 (in Russian).

[30] ROZOVSKII, M.I., *Proceedings of the USSR Academy of Sciences, Mechanics and Engineering*, Vol. 30, 1961 (in Russian).

[31] ERICKSEN, R.H., *Composites*, Vol. 7, 1976, p. 189.

[32] ADAMOVICH, A.G., Short-term Deformation and Long-term Creep of Parapolyamid Yarns, in: *First Conference of Young Specialists in Polymer Mechanics*, Abstracts, Zinatne Press, Riga, 1977, p. 36 (in Russian).

[33] SKUDRA, A.M., BULAVS, F.Ya. and ROCENS, K.A., *Creep and Static Fatigue of Reinforced Plastics*, Zinatne Press, Riga, 1971 (in Russian). German Edition: SKUDRA, A.M., BULAVS, F.J. and ROCENS, K.A., *Kriechen und Zeitstandverhalten von verstärkten Plasten*, VEB Deutscher Verlag für Grundstoffindustrie, Leipzig, 1975.

[34] SKUDRA, A.M. and BULAVS, F.Ya., *Structural Theory of Reinforced Plastics*, Zinatne Press, Riga, 1978 (in Russian).

[35] BULAVS, F.Ya. and RADINSH, I.G., Influence of Matrix Nonlinearity of the Deformational Properties of Composites Under Long-Term Transverse Loading, in: *Mechanics of Composite Materials*, Issue 3, RPI, Riga, 1980, p. 56 (in Russian).

[36] SKUDRA, A.M. (ed.), *Mechanics of Composite Materials*, Issue 3, RPI, Riga, 1980 (in Russian).

[37] BULAVS, F.Ya. and RADINSH, I.G., In-Time State of Stress Variation in Unidirectional Reinforced Plastic Components Under Long-Term Static Loading, in: *Mechanics of Composite Materials*, Issue 3, RPI, Riga, 1980, p. 5 (in Russian).

[38] RADINSH, I.G. and GURVICH, M.R., Viscoelastic Properties of Laminated Reinforced Plastics Under Long-Term In-Plane Loading, in: *Mechanics of Reinforced Plastics*, Issue 4, RPI,

Riga, 1981, p. 76.

[39] GURVICH, M.R., Long-Term Strength of Unidirectional Reinforced Plastics at Failure of Fiber-Matrix Interface, in: *Mechanics of Reinforced Plastics,* Issue 6, RPI, 1983, p. 32 (in Russian).

[40] RABOTNOV, Yu.N., 1970, On Failure of Solids, in: *Strength Problems of a Deformable Solid* (Shipbuilding Press, Leningrad), 1970, p. 353 (in Russian).

[41] TAMUZH, V.P. and KUKSENKO, V.S., *Macromechanics of Failure of Polymeric Materials,* Zinatne Press, Riga, 1978 (in Russian).

[42] GURVICH, M.R., Long-Term Strength of Polymeric Matrix Under Combined Variable State of Stress, in: *Mechanics of Composite Materials,* Issue 5, RPI, Riga, 1982, p. 25 (in Russian).

[43] BULAVS, F.Ya., GURVICH, M.R. and RADINISH, I.G., Long-Term Strength of Unidirectional Reinforced Plastics Mutually Loaded in Tension and Shear, in: *Mechanics of Composite Materials,* Issue 5, RPI, Riga, 1982, p. 33 (in Russian).

[44] BULAVS, F.Ya. and GURVICH, M.R., Continuity Loss in Laminated Plastics Under Long-Term Plane Loading, in: *Mechanics of Composite Materials,* Issue 7, RPI, Riga, 1984, p. 11 (in Russian).

[45] AGARWAL, B.D. and BROUTMAN, L.J., *Analysis of Performance of Fiber Composites,* John Wiley and Sons, New York, 1980.

[46] BAX, J., *Plastics and Polymers,* Vol. 38, 1970, p. 27.

[47] BULAVS, F.Ya. and GURVICH, M.R., Methodics for Rational Designing of Laminated Plastics with Pre-Set Properties of Their Components, in: *Mechanics of Reinforced Plastics,* Issue 6, RPI, Riga, 1983, p. 69 (in Russian).

[48] BULAVS, F.Ya. and GURVICH, M.R., Rational Designing of Laminated Plastics for Long-Term In-Plane Loading, in: *Designing and Optimization of Engineering Constructions,* RPI, Riga, 1983 (in Russian).

[49] SOKOLOV, E.A., Candidate's Dissertation, Institute of Polymer Mechanics, Riga, 1980 (in Russian).

S.T. Mileiko and V.V. Tvardovsky

8

Interaction of macro- and microcracks in a nonhomogeneous solid

8.1 Introduction

A macrocrack in any solid under load gives a birth to microdefects in front of its tip. The scale level of the microdefects can be very different, from the atomic or dislocation level, in which case plasticity of some physical nature will occur, to the rather macroscopic level, microcracks and macrovoids being examples. The fracture toughness of the solid is determined by a complicated interaction of defects on different levels. It appears to be difficult to choose a predominant type of interaction. In this paper, however, we restrict ourselves to the framework of the interaction of the macrocrack, being an initial defect with a system of microcracks created by the macrocrack.

The creation of such microcracks has been observed experimentally in metals [1, 2], ceramics [3, 4], polymers [5, 6] and fibrous composites [7]. To illustrate the situation, Figure 8.1 shows the dependence of microcrack density on the distance from the crack tip [5] and Figure 8.2 shows a typical zone of microcracking in a fibrous composite.

The study of the variety of appearances of a system of microcracks leads to an obvious assumption. Namely, a microdefect such as a microcrack, flaw, etc., should arise not at an arbitrary point but at such a point where an imperfection exists in the structure. The scale level of such imperfections varies from the atomic (vacancies being an example) to quite macroscopic in the case of brittle fibre breaks in composites. In practice, all real materials are nonhomogeneous, so it can be expected that the birth of the microcracks in the vicinity of the crack tip and their coalescence is the universal mode of macrocrack propagation.

The problem of an interaction of the original (macro-) crack with the secondary (micro-) crack was formulated in [8]: only the situation of one or two secondary cracks was considered. The situation with a limited number of microcracks has also bee analyzed [9, 10]. When the number of microcracks is sufficiently large, their overall effect can be described by introducing effective parameters. The mathematics can be simplified by assuming that the size of the microcracking zone is small in comparison with the crack length. An elastic-plastic solid with cracks may then be assumed. The work reported in [11] divides microcracking into a local zone close to the crack tip and an outside zone which behaves effectively as an elastic solid. The stochastic nature of the microcracks in the local zone can be effectively reflected by the behavior of a larger macrocrack. This explains the rise of effective surface energy in relation to the homogeneous surface energy.

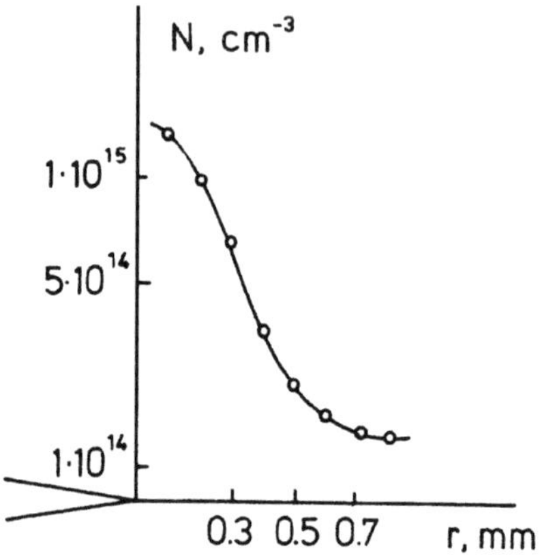

Fig. 8.1: Dependence of microcrack density in a polymer on distance from crack tip.

Fig. 8.2: Fibre breakage near crack surface (Courtesy of V.M. Kiiko).

In contrast, the energy dissipation due to nucleation and growth of the microcracks were not considered in [12]. Only those outside the local region were accounted. Macrocrack propagation was facilitated and the model is somewhat artificial. Criteria such as crack opening displacement, crack tip opening angle, etc., used in conjuction with an elastic-plastic body, are equally inadequate [13]. Unlike the elastic-plastic body, a nonhomogeneous system is often composed of elastic components; it can be characterized by the elastic potential. A Griffith-type criterion can be used to determine the critical load for crack propagation, as was first discussed in [14].

A valid criterion should consider microdefect nucleation and energy dissipation. This has been done for the case of a fibre breaking ahead of a crack [15]. A strong interaction of the neighboring microcracks on the size of the fracture process zone is found. The fracture toughness of a nonhomogeneous ceramic under small scale yielding has been analyzed [16, 17], including the effect of thermal residual stress. It should be clear that interaction of macro- and microcracks cannot be ignored because it can influence the fracture toughness of the solid. Useful and reliable results can only be obtained on the basis of sound assumptions, so that predictions can be verified experimentally.

The objective of this work is to show that errors of several orders of magnitude can result if certain features of macro- and microcracking are ignored. Unavailable analytical solutions are also provided as by-products that are of interest in themselves. For simplicity, the analytical model of longitudinal shear is chosen for an elastic solid containing elastic inclusions. Nonhomogeneity arising from the nucleation of microcracks will be considered. A special feature of the model is that the adhesion at the interfaces of the inclusion and matrix can be random.

To begin with, the delamination of a single inclusion from the matrix will be considered using a Griffith-type criterion. This gives the stress and strain curve for the model of a system with inclusions. The behavior of such a solid is similar to that of a strain-hardened plastic solid. The shape of the stress-strain curve depends upon both the distribution function of the specific energy of the inclusion/matrix interfaces and the boundary conditions. A macrocrack is then introduced into the nonhomogeneous solid. The interaction of microcracks is accounted for by correcting the effective elastic moduli of the solid. A value of the critical crack length and the effective fracture surface energy are found to depend on the distribution function for the interface energy. When the crack length is small in comparison with the characteristic dimension of the delamination zone, the problem can be solved numerically.

8.2 Delamination of inclusion from matrix

To evaluate the behavior of a model made of the cells shown in Figure 8.3, let us first consider the longitudinal shear of a single cell. The interface may fail by the nucleation and Mode III cracking. This will be referred to as microcracking delamination.

Dependence of the delamination stress on the angle of delamination $2\theta_0$ as governed by Equations (8.52) and (8.53) is shown in Figure 8.4. Initially, θ_0 would be small while complete delamination corresponds to the boundary stress reaching some critical value. The criterion of delamination can be obtained from the energy balance

$$\delta A = \delta W + \delta F \tag{8.1}$$

where δA is the work done by the external forces, δW the change of the elastic energy

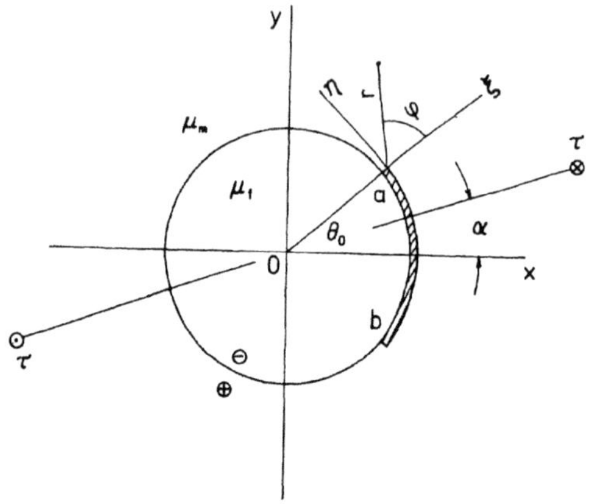

Fig. 8.3: An interface crack.

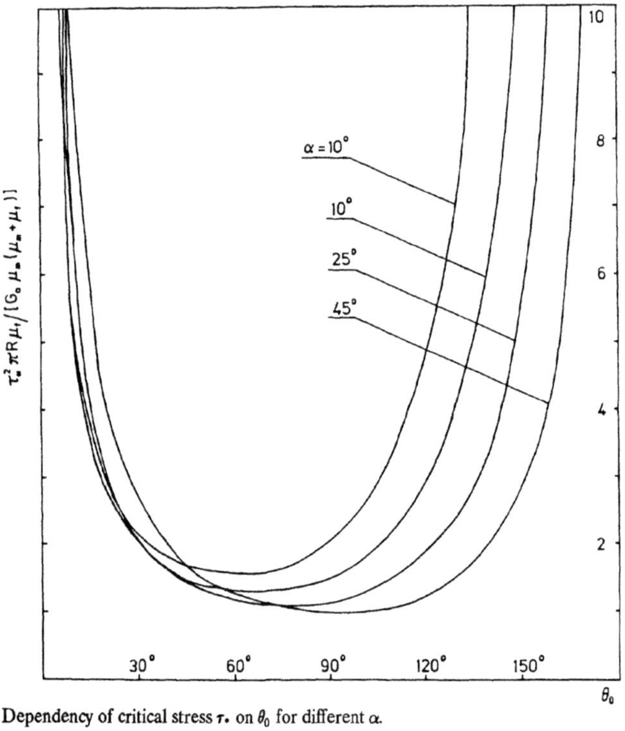

Fig. 8.4: Dependency of critical stress τ_* on θ_0 for different α.

and δF the change of the surface energy.

Let the external forces T_i act on the body with surface S, then

$$
\begin{aligned}
\delta W &= \frac{1}{2}\int_V (\sigma_{ij}^{(1)}\varepsilon_{ij}^{(1)} - \sigma_{ij}^{(0)}\varepsilon_{ij}^{(0)})\,\mathrm{d}V \\
&= \frac{1}{2}\int_S (\sigma_{ij}^{(1)}n_j u_i^{(1)} - \sigma_{ij}^{(0)}n_j u_i^{(0)})\,\mathrm{d}S \\
&= \frac{1}{2}\int_S T_i(u_i^{(1)} - u_i^{(0)})\,\mathrm{d}S = \frac{1}{2}\delta A
\end{aligned}
\tag{8.2}
$$

where $u_i^{(0)}$, $\varepsilon_{ij}^{(0)}$ and $\sigma_{ij}^{(0)}$ are components of the displacement vector, the strain and stress tensors, respectively, in the initial state while $u_i^{(1)}$, $\varepsilon_{ij}^{(1)}$ and $\sigma_{ij}^{(1)}$ are those in the final state, after delamination. Moreover,

$$
\delta W = W_1 - W_0 \tag{8.3}
$$

with W_0 being the elastic energy change due to presence of the bounded inclusion given by Equation (8.48) such that for $\theta_0 = 0$:

$$
W_0 = \frac{\pi R^2}{\mu_m}\frac{\mu_m - \mu_f}{\mu_m + \mu_f}\tau^2. \tag{8.4}
$$

In Equation (8.3), W_1 is the elastic energy change due to complete delamination. For $\mu_f = 0$ or $\theta_0 = \pi$, Equation (8.48) gives

$$
W_1 = \frac{\pi R^2}{\mu_m}\tau^2. \tag{8.5}
$$

Here, τ is the maximum of shear stress at infinity. Let δF be expressed as $2\pi RG$, where $2\pi R$ is the area of the newly formed surface per unit thickness. The effective interface energy G depends on the density, sizes and distribution of microdefects. It corresponds to the work done to fracture the interface and can be measured experimentally. That is,

$$
W_1 - W_0 = 2\pi RG \tag{8.6}
$$

such that

$$
\tau^2 = \frac{\mu_m(\mu_m + \mu_f)}{\mu_f R}G. \tag{8.7}
$$

Alternatively, a thermodynamic analysis can be performed to yield a different form of Equation (8.7):

$$
\tau\gamma = \frac{\mu_m + \mu_f}{\mu_f R}G \tag{8.8}
$$

in which γ is the average shear strain within the area occupied by the fibre.

The effective value of G provides a description for a system with microcracks even though the analysis started with a single interface defect. Since dynamic effects are not included, an underestimate of the critical stresses is anticipated.

8.3 Fibre/matrix delamination and stress-strain response

Consider a composite with circular fibres subjected to longitudinal shear. The total fibre volume fraction is v_f and v_f^d is the volume fraction of fibres which may be delaminated. Let the effective interface energy G be a random variable. The distribution density function for G is $\chi(\omega)$. Then, the volume fraction of fibres after delamination associated with the stress τ is

$$v_\tau^d = v_f^d \int_0^{\omega(\tau)} \chi(\omega)\,\mathrm{d}\omega. \tag{8.9}$$

The delamination of each fibre can be regarded as the creation of a cylindrical hole. The initial effective shear modulus $\mu_0 = \mu_f v_f + \mu_m v_m$ changes to $\mu_1 = \mu_f(v_f - v_f^d) + \mu_m v_m$ and the average stress τ corresponds to

$$\mu = \mu_0 + (\mu_1 - \mu_0)\frac{v_\tau^d}{v_f^d}. \tag{8.10}$$

Approximate expressions for the elastic shear modulus of the composite have been used. The final result will only alter qualitatively if a more exact formula is used.

The delamination criterion of a single fibre in an infinite medium with modulus μ is given by Equation (8.8), i.e.,

$$\tau^2 = \mu(\tau)\frac{\mu_m + \mu_f}{\mu_f R}\,\omega. \tag{8.11}$$

The pair of Equations (8.10) and (8.11) gives the dependence of the average stress τ on the effective shear modulus μ. This gives the $\tau - \gamma$ curve that depends on $\chi(\omega)$, the distribution function for the effective interface energy G. The shape of this curve depends on the strain energy, fracture work, and effective surface energy of the composite.

8.4 Fracture toughness

Let the length of a macrocrack in the composite be $2L$. At infinity, uniform stresses $\tau_{zx} = 0$, $\tau_{zy} = \tau$ are applied as shown in Figure 8.5. The crack surfaces are free from the loads. Fibre/matrix delamination can occur near the crack; this reduces the effective modulus in accordance with Equation (8.10). Assume that

$$L >> R, \quad G_0 >> \omega_0, \quad L >> A^{1/2} \tag{8.12}$$

where A is the area of the delamination zone, G_0 the effective surface energy of a composite without delamination, and ω_0 the value of ω prior to delamination. The second inequality in Equation (8.12) corresponds to a large number of fibres in the delamination area so that the material behavior may be described in terms of the effective parameters. The third inequality implies that $\tau_c << \tau_0$ or $G_0/\omega_0 << L/R$, where τ_c is the critical stress for the crack in a composite with no delamination and τ_0 is the stress at which delamination first occurs, i.e.,

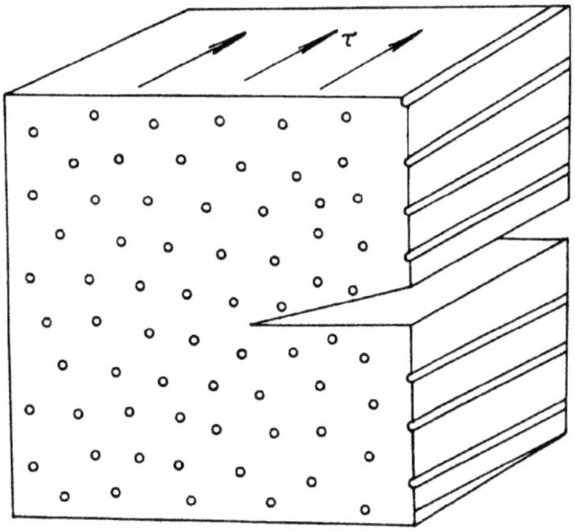

Fig. 8.5: A composite model.

$$\tau_c^2 = \frac{2\mu_0 G_0}{\pi L}; \quad \tau_0^2 = \frac{\mu_0(\mu_m + \mu_f)}{\mu_f R}\omega_0. \tag{8.13}$$

Making use of the solution for the elastic-plastic problem of a solid with strain hardening [18], the radius of a quasi-plastic zone or the zone of delamination as shown in Figure 8.6 is found:

$$\rho = \frac{1}{2}L_1\left(\frac{\tau}{\tau_0}\right)^2. \tag{8.14}$$

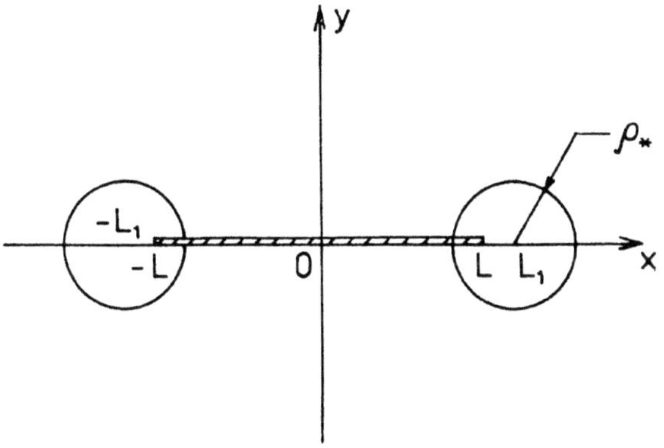

Fig. 8.6: Delamination zones around crack tips.

The centers of these circular zones are located at $(\pm L_1, 0)$ such that

$$L_1 = L\left(1 - \frac{\varepsilon\tau^2}{\tau_0^2}\right)^{-1}$$

$$\varepsilon = \tau_0\gamma_0 \int_{\gamma_0}^{\infty} \frac{\mathrm{d}\gamma}{\gamma^2\tau(\gamma)} - \frac{1}{2} \tag{8.15}$$

$$\gamma_0 = \frac{\tau_0}{\mu_0}.$$

To reiterate, the function $\tau(\gamma)$ depends on the distribution function $\chi(\omega)$.

The stress-strain state in an elastic region can be found from the function $\phi(\varsigma)$ where $\varsigma = x + iy$. It is holomorphic everywhere except the points on the cut that corresponds to the macrocrack:

$$\tau_{zx} - i\tau_{zy} = \phi'(\varsigma),$$

$$2\mu_0\mu_z = \phi(\varsigma) + \overline{\phi(\varsigma)}. \tag{8.16}$$

The boundary conditions at infinity and at the boundary of quasiplastic zone suggest that

$$\phi'(\varsigma) = -\frac{i\tau\varsigma}{\sqrt{\varsigma^2 - L_1^2}}. \tag{8.17}$$

Since energy dissipation due to friction has been neglected, unloading in the $(\tau - \gamma)$ domain will take place along a straight line toward the origin; the behavior will be quasielastic. The compliance of the system can decrease. With the change in elastic energy due to the cut given by $W_{00} = \pi L^2\tau^2/2\mu_0$, the total change in elastic energy including the two quasiplastic zones is given by

$$W_0 = \frac{\pi L_1^2\tau^2}{2\mu_0}. \tag{8.18}$$

The zones of delamination are much smaller than the crack length so that no interaction occurs. In other words, the radius of the delamination zone is assumed to be independent of the crack length and it can be regarded as a material parameter.

The behavior of a composite is different from that of an elastic-plastic solid. Nonelasticity in a component is a consequence of nucleation of microcracks accompanied by energy dissipation; this changes the elastic moduli. Assuming that the size of the delamination zone is constant during crack growth, the change in elastic energy due to unloading of the material behind the crack can be computed. It can be shown that this effect is negligibly small and does not seriously affect the total energy balance. It will therefore not be included, and a constant fracture process zone is assumed.

Let ρ_* be the critical radius of the quasiplastic zone referred to fiber delamination with the average interface energy $\langle\omega\rangle$. It defines the effective area $A^* = \pi\rho_*^2$ and the energy dissipated due to creation of the quasiplastic zone $G_A \cdot A^*$. Here, $G_A = 2v_f^d\langle\omega\rangle/R$. Shear strain along the line corresponding to $\rho = \rho_*$ is given by

$$\langle\gamma\rangle^2 = \frac{\mu_m + \mu_f}{\mu_f\mu(\langle\omega\rangle)R}\langle\omega\rangle. \tag{8.19}$$

Hence, Equation (8.14) yields

$$\tau_* = \tau_0\sqrt{\frac{2\rho_*}{\kappa L_1}}; \quad \kappa = \frac{\gamma_0\tau_0}{\langle\gamma\rangle\tau(\langle\gamma\rangle)}. \tag{8.20}$$

The Griffith-type energy balance gives

$$\delta W = 2G_0\delta L + 2G_A\delta A^*. \tag{8.21}$$

Equations (8.15) and (8.21) lead to

$$\delta W = \pi L\tau^2\left(1 - \frac{\varepsilon\tau^2}{\tau_0^2}\right)^{-2}\delta L. \tag{8.22}$$

The condition that ρ_* is constant gives

$$\delta A^* = 2\rho_*\delta L. \tag{8.23}$$

Therefore, Equation (8.21) yields

$$\tau_* = \tau_c\frac{L}{L_1}\sqrt{1 + 4\rho_*\frac{\langle\omega\rangle - v_f^d}{G_0R}}. \tag{8.24}$$

Finally, making use of Equation (8.20), the following results are obtained

$$\tau_*^2 = \frac{2\mu_0(\mu_m + \mu_f)\omega_0\rho_*}{\pi L\kappa\mu_f R}$$

$$\rho_* = R\kappa\left(\frac{\mu_m + \mu_f}{\mu_f G_0}\omega_0 - \frac{4v_f^d\omega_0}{\pi G_0} + \frac{4\varepsilon R}{\pi L}\right)^{-1} \tag{8.25}$$

$$G_* = G_0\left(1 - \frac{4v_f^d\mu_f}{\pi(\mu_m + \mu_f)} + \frac{4\varepsilon\mu_f R G_0}{\pi(\mu_m + \mu_f)\omega_0 L}\right)^{-1}.$$

The above results are valid for a broad class of functions $\chi(\omega)$ provided that the delamination zone is small. Because of the difficulty of analyzing the fracture process for arbitrary $\chi(\omega)$, a bimodal symmetry model, given in Figure 8.7, will be considered to study the effect of dispersion on the effective surface energy. A mean value of ω or $\langle\omega\rangle$ is taken to be constant. The peaks at ω_1 and ω_2 characterize the dispersion while the areas s_1 and s_2 determine the symmetry of the third moment.

It can be seen from Equations (8.10) and (8.11) that the distribution function determines the stress-strain curve $\tau(\gamma)$ or vice versa. An alternative form of the distribution function is shown in Figure 8.8 and characterized by

$$\chi(\omega) = 0, \quad 0 < \omega < \omega_{10}, \quad \omega_{11} < \omega < \omega_{20}, \quad \omega > \omega_{21};$$

$$\chi(\omega) = \frac{\Gamma\tau_1^2}{\omega^2}, \quad \omega_{10} < \omega < \omega_{11};$$

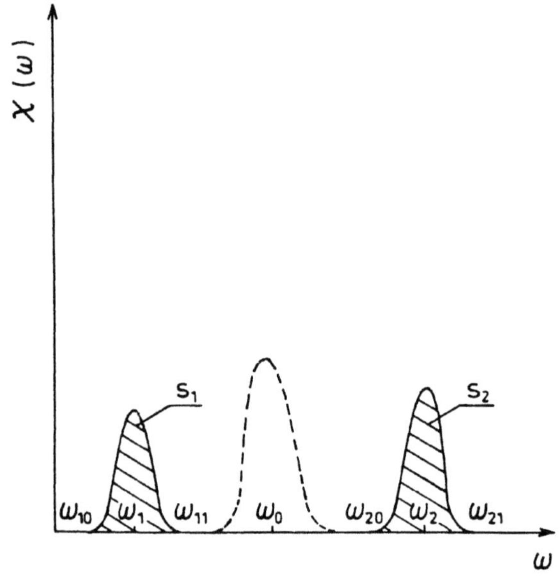

Fig. 8.7: Bimodular distribution function.

$$\chi(\omega) = \frac{\Gamma\tau_2^2}{\omega^2}, \quad \omega_{20} < \omega < \omega_{21};$$

$$\Gamma = \frac{\mu_f R}{[(\mu_m + \mu_f)(\mu_0 - \mu_1)]} \tag{8.26}$$

which correspond to the stress-strain curve with two yield stresses. The values ω_{10}, ω_{11}, ω_{20} and ω_{21} can be expressed via the values τ_1, τ_2, μ_0, μ_1 and μ_s by means of Equation (8.11). Here, $\mu_s = \mu_0 + (\mu_1 - \mu_0)s$ and $s = s_1/(s_1 + s_2)\sigma$. The stress-strain curve with a single yield point τ_0 can be used for comparison. It corresponds to the dotted curve in Figure 8.7.

The crack problem solution of a strain-hardening solid [18] for antiplane shear leads to

$$A^* = L^2 \cdot f_A\left(\frac{\tau}{\tau_1}\right)$$

$$W = \frac{1}{2}\pi L^2\tau^2 f_w\left(\frac{\tau}{\tau_1}\right)/\mu_0. \tag{8.27}$$

Numerical differentiation with respect to crack length was not necessary in order to use the Griffith approach. The functions $f_A(\tau/\tau_1)$, $f_W(\tau/\tau_1)$ and the size of the delamination zone are obtained numerically. With $G_A = 2v_f^d \langle\omega\rangle/R$ being a constant that does not depend on a distribution function, Equation (8.21) can be generalized for

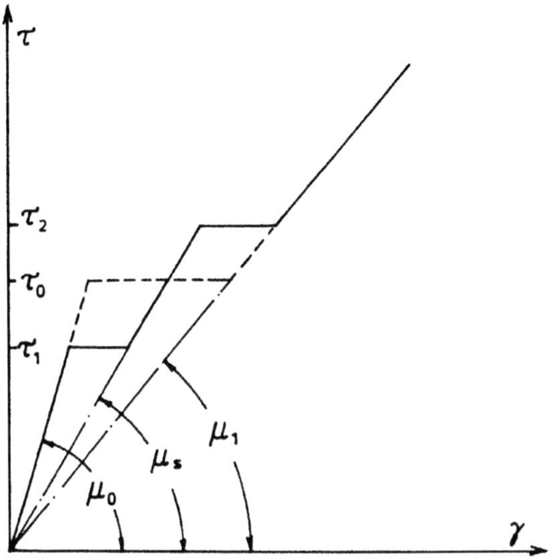

Fig. 8.8: Stress-strain curves with unimodular (dotted line) and bimodular distribution function.

the quasiplastic zones.

Equation (8.21) for a fixed value of τ can be solved with respect to L. Figure 8.9 gives the dependence of the normalized stress $\sigma = \tau/\tau_0$ on the normalized critical crack length $L_* = L\pi\tau_0^2/(2\mu_0 G_0)$. Knowing that $\tau_0^2 = \mu_0(\mu_f + \mu_m)\omega_0/(\mu_f R)$, the expression $L_* = L\pi(\mu_m + \mu_f)\omega_0/(2RG_0\mu_f)$ follows. The curves in Figure 8.9 are labelled from 1, 2, ..., 9; they correspond to $v_f = 0.55$, $v_f^d = 0.44$ and $\mu_f/\mu_m = 50.0$ in Table 8.1. The critical stress depends strongly on the asymmetry of the distribution function, i.e., the parameter s, and even more strongly on the dispersion, i.e., the parameter τ_1/τ_0.

8.5 Conclusion

The energy release rate and the effective surface energy in a composite with nonhomogeneous interfaces depend not only on the interface energy but also on both the second and the third moments of the distribution function. Material parameters are assumed to be stochastic in nature so that they can influence the effective fracture characteristics. In some cases, the second and higher moments of a distribution function can effect these properties.

The effective surface energy is not characteristic of the composite material because of its dependence on the crack length. This dependence can be neglected when the crack is large in comparison with the delamination zone. Hence, linear fracture mechanics can be applicable to large cracks.

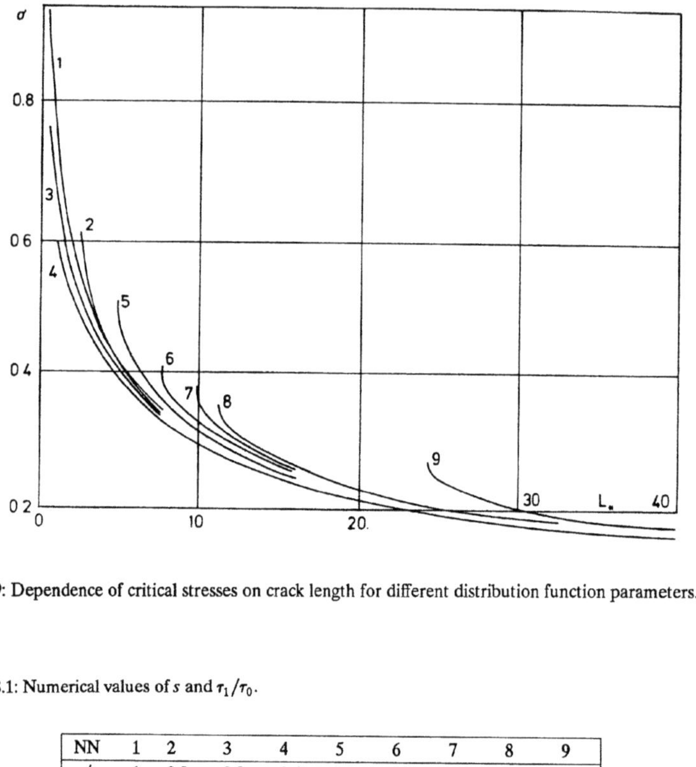

Fig. 8.9: Dependence of critical stresses on crack length for different distribution function parameters.

Table 8.1: Numerical values of s and τ_1/τ_0.

NN	1	2	3	4	5	6	7	8	9
τ_1/τ_0	1	0.8	0.8	0.6	0.8	0.8	0.6	0.8	0.6
s	-	0.70	0.85	0.85	0.55	0.40	0.70	0.25	0.55

8.6 Appendix: Matrix/fibre interface fracture under longitudinal shear

Elastostatic problems of cracks between dissimilar media have been studied previously [19 – 22]. Conceptual difficulties arise because of the oscillatory singularities and mutual penetration of the crack tip surfaces. Attempts have been made to eliminate such singularities [23]. Interface cracks in antiplane shear do not possess oscillatory singularities. The solution for a single crack and a pair of symmetrically located cracks on the interface between a matrix and a circular inclusion has been obtained [24].

Problems involving inclusions of various shapes have been considered in [25 – 27]. An approximate solution for a periodically spaced elastic inclusion with interface cracks has been given [28]. Singular problems for semi-infinite interface cracks are also solved [29].

In this section, the longitudinal shear problem for an arbitrary array of cracks at the interface between a matrix and a circular fibre is solved. A constant load is applied at infinity while tractions or displacements may be specified on the cracks surfaces. The case of a single interface crack is presented in detail.

The change in elastic energy due to the crack and inclusion are calculated. Fracture criteria of the Griffith and Irwin type are applied. They yield a stress intensity factor K_* for the interface crack. Possible crack growth from the interface is also discussed.

Let a circular inclusion of unit radius be placed in an infinite solid. The shear moduli of the two solids are μ_f and μ_m. The cracks are located along $l_0(t)$, while $l_1(t)$ denotes the bonded portion. Tractions $f^+(t)$ and $f^-(t)$ with $t = e^{i\theta}$ are presecribed on $l_0(t)$ and continuity of tractions and displacements is satisfied on $l_1(t)$. Constant stresses τ_{zx}^∞ and τ_{zy}^∞ are applied at infinity.

The stress and displacement fields for longitudinal shear can be expressed in terms of a single holomorphic function $\phi(\varsigma)$ [30]:

$$\tau_{zx} - i\tau_{zy} = \phi'(\varsigma) \tag{8.28a}$$

$$2\mu u_z = \phi(\varsigma) + \overline{\phi(\varsigma)} \tag{8.28b}$$

where $\varsigma = x + iy$. Equation (8.28b) may be expressed as

$$2\mu\frac{\partial u_z}{\partial\theta} = i\varsigma\phi'(\varsigma) + \overline{i\varsigma\phi'(\varsigma)}. \tag{8.29}$$

The stress at an element with the normal $n:(l,\ m)$ can be written as

$$\tau_n = l\tau_{zx} + m\tau_{zy}. \tag{8.30}$$

The condition at the interface is

$$2\tau_n = t\phi'(t) + \overline{t\phi'(t)}. \tag{8.31}$$

The quality

$$l + im = \cos\theta + i\sin\theta = t \tag{8.32}$$

has been used to obtain Equation (8.31). The boundary conditions on $l_0(t)$ are

$$\tau_n^+(t) = f^+(t), \quad \tau_n^-(t) = f^-(t). \tag{8.33}$$

Continuity on $l_1(t)$ implies that

$$\tau_n^+(t) = \tau_n^-(t), \quad \frac{\partial u_2^+}{\partial\theta} = \frac{\partial u_2^-}{\partial\theta}. \tag{8.34}$$

Conditions at infinity can be expressed as

$$\phi'(\varsigma) = T + N\cdot\varsigma^{-1} + O(\varsigma^{-2}) \tag{8.35}$$

where $T = \tau_{zx}^\infty - i\tau_{zy}^\infty$ and $2\pi N$ is the total force applied to the crack surfaces.

Let the holomorphic functions $\phi_f'(\varsigma)$ and $\phi_m'(\varsigma)$ refer to the inside and to the outside of the circular inclusion, respectively. The function $F_1(\varsigma)$ is holomorphic everywhere except at a finite number of points:

$$F_1(\varsigma) = \frac{1}{2}\begin{cases} \varsigma\phi_f'(\varsigma) + \dfrac{1}{\varsigma}\overline{\Phi}_m'\left(\dfrac{1}{\varsigma}\right) & |\varsigma| < 1 \\ \varsigma\phi_m'(\varsigma) + \dfrac{1}{\varsigma}\overline{\Phi}_f'\left(\dfrac{1}{\varsigma}\right), & |\varsigma| > 1\text{'} \end{cases} \tag{8.36}$$

Construct the function $F_2(\varsigma)$ such that

$$F_2(\varsigma) = \frac{1}{2}\begin{cases} \varsigma\phi_f'(\varsigma) - \dfrac{1}{\varsigma}\overline{\Phi}_m'\left(\dfrac{1}{\varsigma}\right), & |\varsigma| < 1 \\ \varsigma\phi_m'(\varsigma) - \dfrac{1}{\varsigma}\overline{\Phi}_f'\left(\dfrac{1}{\varsigma}\right), & |\varsigma| > 1\text{'} \end{cases} \tag{8.37}$$

If

$$F_i(\varsigma) = \begin{cases} F_i^-(\varsigma), & |\varsigma| < 1 \\ F_i^+(\varsigma), & |\varsigma| > 1 \end{cases} \tag{8.38}$$

then $F_1(\varsigma)$ and $F_2(\varsigma)$ are connected:

$$F_1^-(\varsigma) + F_2^-(\varsigma) = \overline{F}_1^+\left(\frac{1}{\varsigma}\right) - \overline{F}_2^+\left(\frac{1}{\varsigma}\right),$$

$$\overline{F}_1^-\left(\frac{1}{\varsigma}\right) - \overline{F}_2^-\left(\frac{1}{\varsigma}\right) = F_1^+(\varsigma) + F_2^+(\varsigma). \tag{8.39}$$

At the boundary $l(t) = l_0(t) + l_1(t)$, the following relations hold:

$$F_1^+(t) = \overline{F_1^-(t)}, \quad F_2^+(t) = -\overline{F_2^-(t)}. \tag{8.4}$$

In accordance with Equations (8.36) and (8.37), the functions $\phi_f'(\varsigma)$ and $\phi_m'(\varsigma)$ can be expressed by $F_1(\varsigma)$ and $F_2(\varsigma)$:

$$\phi_f'(\varsigma) = \frac{1}{\varsigma}(F_1^-(\varsigma) + F_2^-(\varsigma))$$

$$\phi_m'(\varsigma) = \frac{1}{\varsigma}(F_2^+(\varsigma) + F_2^+(\varsigma)). \tag{8.41}$$

Making use of Equations (8.29), (8.34), (8.40) and (8.41), it is possible to present the boundary conditions in terms of the sectionally holomorphic functions $F_1(\varsigma)$ and $F_2(\varsigma)$:

$$\begin{cases} F_1^+(t) - F_1^-(t) = 2\dfrac{\mu_m - \mu_f}{\mu_m + \mu_f} F_2^+(t), & t \in l_1(t) \\ F_2^+(t) + F_1^-(t) = f^+(t) + f^-(t), & t \in l_0(t) \end{cases} \tag{8.42}$$

$$\begin{cases} F_2^+(t) - F_2^-(t) = 0, & t \in l_1(t) \\ F_2^+(t) - F_2^-(t) = f^+(t) - f^-(t), & t \in l_0(t) \end{cases} \tag{8.43}$$

Solutions to the Hilbert–Riemann problem are well known [31, 32]. The functions that satisfy Equations (8.42) and (8.43) can be written in terms of the Cauchy singular integrals

$$F_2(\varsigma) = -\frac{1}{2\pi i}\int_{l_0} \frac{f^+(t) - f^-(t)}{t - \varsigma}\,\mathrm{d}t + R_2(\varsigma)$$

$$F_1(\varsigma) = -\frac{X(\varsigma)}{2\pi i}\left\{\int_{l_0} \frac{\mu_m - \mu_f}{\mu_m + \mu_f}\frac{F_2^+(t)\,\mathrm{d}t}{X^+(t)(t - \varsigma)} + \right.$$

$$\left. + \int_{l_1} \frac{f^+(t) + f^-(t)}{X^+(t)(t - \varsigma)}\,\mathrm{d}t\right\} + X(\varsigma)R_1(\varsigma) \tag{8.44}$$

where $X(\varsigma)$ is the homogeneous solution of Equation (8.42). The rational functions $R_1(\varsigma)$ and $R_2(\varsigma)$ are completely determined by the conditions at infinity and the net force applied to the crack surfaces.

The second fundamental problem of elasticity deals with specifying the displacements on the crack surfaces:

$$u_z^+(t) = v_+(t); \quad u_z^-(t) = v_-(t) \quad \text{for} \quad t \in l_0(t). \tag{8.45}$$

The boundary conditions on $l_1(t)$ are the same as that in the first problem. On $l_0(t)$, there results

$$F_1^+(t) - F_1^-(t) = t(\mu_m v_+'(t) - \mu_f v_-'(t))$$
$$F_2^+(t) + F_2^-(t) = t(\mu_m v_+'(t) + \mu_f v_-'(t)). \tag{8.46}$$

Similar expressions for $F_1(\varsigma)$ and $F_2(\varsigma)$ can be written.

For a single interface crack situated along an arc of radius R symmetrically with respect to the x-axis as shown in Figure 8.3, it is found that

$$\phi_f'(\varsigma) = \phi_0'(\varsigma) + \frac{\mu_f}{\mu_m + \mu_f}\left(T - \frac{\overline{T}R^2}{\varsigma^2}\right)$$

$$\phi_m'(\varsigma) = \phi_0'(\varsigma) + \frac{\mu_m}{\mu_m + \mu_f}\left(T - \frac{\overline{T}R^2}{\varsigma^2}\right) \tag{8.47}$$

where

$$\tau_n^{\pm}(t) = 0, \quad t \in l_0(t)$$

and

$$\phi_0'(\varsigma) = \frac{\mu_f}{\mu_m + \mu_f} \frac{T(\varsigma - R\cos\theta_0)\varsigma^2 - \overline{T}R^2(R - \varsigma\cos\theta_0)}{\varsigma^2\sqrt{\varsigma^2 - 2R\varsigma\cos\theta_0 + R^2}}.$$

It can be shown that the change in elastic energy is given by

$$W_0 = \frac{\pi R^2}{\mu_m}\left\{\frac{\mu_m - \mu_f\cos\theta_0}{\mu_m + \mu_f} T\overline{T} + \frac{1}{2}\frac{\mu_f \sin^2\theta_0}{\mu_m + \mu_f} Re(T^2)\right\}. \tag{8.48}$$

For symmetrical loading ($\tau_{zy}^{\infty} = 0$), the Griffith's condition

$$\frac{\partial W_0}{\partial \theta_0} = 2RG_a \tag{8.49}$$

gives the critical stress τ_*:

$$\tau_*^2 = \frac{2G_a\mu_m(\mu_m + \mu_f)}{\pi R\mu_f \sin\theta_0(1 + \cos\theta_0)}. \tag{8.50}$$

For nonsymmetrical loading ($\tau_{zy}^{\infty} \neq 0$), fracture would start at one of the two crack tips. The Griffith's criterion now gives

$$\frac{\mathrm{d}W_0}{\mathrm{d}\theta_0} = RG_a. \tag{8.51}$$

The symmetry of the system changes as one crack tip moves. With $\tau_{zx}^{\infty} = \tau\cos\alpha$ and $\tau_{zy}^{\infty} = \tau\sin\alpha$, Equation (8.50) takes the form

$$\tau_*^2 = \frac{G_a\mu_m(\mu_m + \mu_f)}{\pi R\mu_f \sin\theta_0 \cos^2(\theta_0/2 - \alpha)} \tag{8.52}$$

if $\cos^2(\theta_0/2 - \alpha) > \cos^2(\theta_0/2 + \alpha)$. Otherwise,

$$\tau_*^2 = \frac{G_a\mu_m(\mu_m + \mu_f)}{\pi R\mu_f \sin\theta_0 \cos^2(\theta_0/2 + \alpha)}. \tag{8.53}$$

Introduce the local coordinate system (ξ, η) as defined in Figure 8.3 with

$$\varsigma = \mathrm{RE}^{\pm i\theta_0}\left\{1 + \frac{r}{R}e^{i\phi}\right\}.$$

Here, r and ϕ are the local polar coordinates. The plus sign refers to the crack tip a and the minus to b. Using Equation (8.47), the stresses at the vicinity of the crack tips are found:

$$\tau_{\xi z}^{a} = \frac{\mu_f}{\mu_f + \mu_m}\sqrt{\frac{2R\sin\theta_0}{r}}\left\{\tau_x^{\infty}\cos\frac{\theta_0}{2} + \tau_{zy}^{\infty}\sin\frac{\theta_0}{2}\right\}\cos\left\{\frac{\pi}{4} - \frac{\phi}{2}\right\}$$

$$\tau^a_{\eta z} = -\frac{\mu_f}{\mu_f + \mu_m}\,\frac{2R\sin\theta_0}{r}\left(\tau^\infty_{zx}\cos\frac{\theta_0}{2} + \tau^\infty_{zy}\sin\frac{\theta_0}{2}\right)\sin\left(\frac{\pi}{4} - \frac{\phi}{2}\right)$$

$$\tau^b_{\xi z} = \frac{\mu_f}{\mu_m + \mu_f}\,\frac{2R\sin\theta_0}{r}\left(\tau^\infty_{zx}\cos\frac{\theta_0}{2} - \tau^\infty_{zy}\sin\frac{\theta_0}{2}\right)\cos\left(\frac{\tau}{4} + \frac{\phi}{2}\right) \tag{8.54}$$

$$\tau^b_{\eta z} = \frac{\mu_f}{\mu_m + \mu_f}\,\frac{2R\sin\theta_0}{r}\left(\tau^\infty_{zx}\cos\frac{\theta_0}{2} - \tau^\infty_{zy}\sin\frac{\theta_0}{2}\right)\sin\left(\frac{\tau}{4} + \frac{\phi}{2}\right).$$

The shear stresses reach a maximum on tangential planes at points a and b as given by

$$\tau^a = \frac{\mu_f}{\mu_m + \mu_f}\,\frac{2R\sin\theta_0}{r}\cos\left(\frac{\theta_0}{2} - \alpha\right)\tau$$

$$\tau^b = \frac{\mu_f}{\mu_m + \mu_f}\,\frac{2R\sin\theta_0}{r}\cos\left(\frac{\theta_0}{2} + \alpha\right)\tau. \tag{8.55}$$

Equations (8.54) and (8.55) can be written in a usual form using the stress intensity factor K where the subscript III will be henceforth be omitted.

The interface energy G_a and the stress intensity factor K_* can be related. Using the well-known relation

$$G_a = \lim_{\delta\to 0}\frac{1}{2\delta}\int_0^\delta \tau^a_{\xi z}\left(r, \frac{\tau}{2}\right)\left[u_z\left(\delta - r, \frac{3}{2}\tau\right) - u_z\left(\delta - r, -\frac{\tau}{2}\right)\right]dr \tag{8.56}$$

and making use of

$$\tau^a_{\xi z}\left(r, \frac{\pi}{2}\right) = \frac{K}{\sqrt{2\pi r}} + O\left(\sqrt{\frac{r}{R}}\right),$$

$$u_z\left(r, -\frac{\pi}{2}\right) = u_z(a) - \frac{K}{\mu_m}\sqrt{\frac{2r}{\pi}} + O\left[\left(\frac{r}{R}\right)^{3/2}\right] \tag{8.57}$$

$$u_z\left(r, \frac{3}{2}\pi\right) = u_z(a) + \frac{K}{\mu_f}\sqrt{\frac{2r}{\pi}} + O\left[\left(\frac{r}{R}\right)^{3/2}\right]$$

it is found that

$$K_* = \sqrt{\frac{4G_a\mu_m\mu_f}{\mu_m + \mu_f}}. \tag{8.58}$$

The interface crack will propagate along the interface providing that

$$K_* < K^*_m,\quad K_* < K^*_f \tag{8.59}$$

in which K^*_m and K^*_f are the critical stress intensity factors for the matrix and the inclusion, respectively, i.e.

$$K^*_m = \sqrt{2G_m\mu_m},\quad K^*_f = \sqrt{2G_f\mu_f}. \tag{8.60}$$

The crack will propagate along the tangential line into the matrix if

$$K_* \geq K_m^*, \quad K_* < K_f^* \tag{8.61}$$

which is valid for quasistatic crack propagation only. There remains a dynamical consideration. In that case, the interface crack may run into either the matrix or inclusion even if Equation (8.59) were satisfied.

The Irwin criterion

$$K = K_* \tag{8.62}$$

can also applied to an interface crack. Using Equations (8.58) and (8.55), Equation (8.52) applied to the crack tip '*a*' and (8.53) to '*b*'. Therefore, the Griffith criterion in Equation (8.51) and Irwin's criterion given by (8.62) are equivalent for interface cracks.

The dependences of τ_* on θ_0 given by (8.52) and (8.53) are shown in Figure 8.4. A crack with angular length $2\theta_0 < 2\theta_*$ propagate in an unstable manner until it reaches the condition $2\theta_2 > 2\theta_*$. It then stops and propagates in a stable quasistatic manner as the load increases. If the angular $2\theta_0 \geq 2\theta_1$, then the crack length grows with increasing load in a stable quasistatic fashion. The situation depends on a particular combination of the values θ_0 and α. Note, in particular, that the value of θ_* is a solution of the equation

$$\cos\theta + \cos 2(\theta + \alpha) = 0, \tag{8.63}$$

which lies within the interval $(0, \pi)$.

References

[1] CRUSSARD, C., PLATEAU, J., TAMHANCAR, R., HENRY, G. and LAJEUNESSE, D., in *Fracture,* edited by Averbach *et al.,* Wiley, New York, 1959, p. 524.

[2] HOAGLAND, R.A., ROSENFIELD, A.R. and HAHN, G.T., *Met. Trans.,* Vol. 3, No. 2, 1972, p. 123.

[3] CLAUSSEN, N., *J. Amer. Ceram. Soc.,* Vol. 59, No. 1–2, 1976, p. 49.

[4] CLAUSSEN, N., STEEB, J. and PABST, R.F., *Amer. Ceram. Soc. Bull.,* Vol. 56, No. 6, 1977, p. 559.

[5] ZHURKOV, S.N., KUSENKO, V.S. and SLUTSKER, A.I., *Fizika Tverdogo Tela (Solid State Physics),* Vol. 11, No. 2, 1969, p. 296 (in Russian).

[6] LEKSOVSKY, A.M., BASKIN, B.L., GOREUBERG, A.Ja., USAMOV, G.H. and REGEL, V.R., *Fizika Tverdogo Tela (Solid State Physics),* Vol. 25, No. 4, 1983, p. 1096 (in Russian).

[7] MILEIKO, S.T., SOROKIN, N.M. and ZIRLIN, A.M., *Mekhanica Compozitnyh Materialov,* No. 6, 1976, p. 1010 (in Russian).

[8] COOK. J. and GORDON, J.E., *Proc. Roy. Soc.,* Vol. A–282, No. 1391, 1964, p. 508.

[9] FU, Y. and EVANS, A., *Acta. Met.,* Vol. 30, No. 8, 1982, p. 1619.

[10] MILEIKO, S.T. and RABOTNOV, Yu.N., *Uspekhi Mekhanici (Advances in Mechanics),* Vol. 3, No. 1, 1980, p. 3.

[11] EVANS, A.G., HEUER, A.H. and PORTER, D.L., in *Proc. Fourth Inter. Conf. in Fracture,* Waterloo, Canada, edited by Taplin, Vol. 1, 1977, p. 529.

[12] ROMALIS, M.B. and TAMUZH, V.P., *Mekhanica Compozitnyh Materialov,* No. 1, 1984, p. 42 (in Russian).

[13] SMITH, E., *J. Mater. Sci. Lett.,* Vol. 2, No. 5, 1983, p. 204.

[14] GLUCKLICH, J. and COHEN, L.J., *Int. J. Fract. Mech.*, Vol. 3, 1967, p. 278.
[15] MILEIKO, S.T. and SULEIMANOV, F.H., *Mekhanica Compozitnyh Materialov,* No. 3, 1981, p. 421 (in Russian).
[16] POMPE, W., *Neue Hutte,* Vol. 23, No. 9, 1978, p. 321.
[17] KREHER, W. and POMPE, W., *J. Mat. Sci.,* Vol. 16, 1981, p. 694.
[18] RICE, J.R., *Trans. ASME, Ser. E, J. Appl. Mech.,* Vol. 34, No. 2, 1967, p. 287.
[19] CHEREPANOV, G.P., *Izv. Akad. Nauk SSSR, Otdel Tekhnickeshik Nauk, Mechnica i Mashinostroenie,* No. 1, 1962, p. 131 (in Russian).
[20] ERDOGAN, F., *Trans. ASME, Ser. E, J. Appl. Mech.,* Vol. 30, No. 2, 1963, p. 232.
[21] ENGLAND, A.H., *Trans. ASME, Ser. E, J. Appl. Mech.,* Vol. 32, No. 2, 1965, p. 400.
[22] RICE, J.R. and SIH, G.C., *Trans. ASME, Ser. E, J. Appl. Mech.,* Vol. 32, No. 2, 1965, p. 418.
[23] COMNINOU, M. and DUNDURS, J., The interface crack in retrospect, in: *Proc. First USA - USSR Symp. on Fracture of Comp. Materials,* Riga, edited by G.C. Sih and V.P. Tamuzs, The Netherlands, 1979.
[24] SMITH, E., *Int. J. Eng. Sci.,* Vol. 7, No. 9, 1969, p. 973.
[25] SENDECKYJ, G.P., *Eng. Fract. Mech.,* Vol. 6, No. 1, 1974, p. 33.
[26] NUISMER, R.J. and SENDECKYJ, G.P., *Trans. ASME, Ser. E, J. Appl. Mech.,* Vol. 44, 1977, p. 625.
[27] BEREZHNITSKY, L.T., PANASYUK, V.V. and SAKHNENKO, A.M., *Fisiko-Khimicheskaja Mechanica Materialov,* Vol. 17, No. 5, 1981, p. 67 (in Russian).
[28] VANIN, G.A., *Prikladnaja Mekhanica (Applied Mechanics),* Vol. XIII, No. 8, 1977, p. 35 (in Russian).
[29] KULIEV, V.D., *Doklady Akad. Nauk Az. SSR,* No. 9, 1979, p. 17 (in Russian).
[30] BARENBLATT, G.I. and CHEREPANOV, G.P., *Prikladnaja Matematika i Mechanics (Applied Mathematics and Mechanics),* Vol. 25, NO. 6, 1961, p. 1110 (in Russian).
[31] GAKHOV, F.D., *Kraevye Zadachi,* Nauka, Moscow, 1977 (in Russian).
[32] MUSKHELSISHVILI, N.I., *Singular Integral Equations,* P. Noordhoff, Groningen, 1963.

A.N. Romanov and M.M. Gadenin

9

Energy balance for elastoplastic fracture: static and cyclic loading

9.1 Introduction

Fracture criteria will be used in fatigue life estimation involving both crack initiation and propagation. The consideration of static fracture work [1] alone may not be sufficient because experiments [2] have shown that the energy spent in the cyclic deformation of a material can be much larger than the static fracture energy. Moreover, energy dissipated in the form of heat [3] can be a large portion of the total energy under both static and cyclic loading.

Past works in fracture mechanics have made use of the specific surface energy [4], the crack propagation energy G [5] and the J-integral [6]. When both elastic and plastic deformation are present in the cracked specimens, however, the governing parameters become load-history dependent; experimental guidance is needed to develop the corresponding crack growth criteria because not all of the energy will be spent on surface creation. Energy dissipated by heat must also be accounted for.

9.2 Energy balance

When a material is deformed, the energy within the system must be conserved such that [3, 7]

$$A = E + Q \tag{9.1}$$

in which A is the mechanical work, E the stored energy and Q the thermal energy. Quantitative assessment of Equation (9.1) requires a clear distinction between the thermal energy given off as dissipation and the mechanical energy used in the deformation process. In [8] a method is proposed for evaluating these quantities in static and cyclic loading.

Method of measurement

The mechanical work used in the deformation process is determined from the area under the two-coordinate plots from a recorder for both static and cyclic load. Dynamometer and extensometer are used. Thermal energy during deformation is evaluated by measuring the temperature rise as a result of loading. The measurements are conducted in vacuum using chrome-copper thermocouples; the response is amplified to achieve an accuracy of $\pm 10^{-2}$ °C.

Thermal energy

Suppose that Q is the total thermal energy. Let Q_T denote the portion of heat lost by conduction through the specimen grips; Q_P the amount generated in heating up the specimen; and Q_T is the steady state transfer. The heat lost due to radiation Q_n is negligibly small because the specimen is heated up to only one-tenth for a degree. This was estimated by using a heat-absorbing screen placed next to the specimen. Convective heat transfer Q_K is also neglected because the experiment is conducted in a vacuum of 10^{-3} mm Hg. In this respect, the total thermal energy equation becomes

$$Q = Q_T + Q_P + Q_n + Q_k \simeq Q_T + Q_P. \tag{9.2}$$

9.3 Heat conduction and generation

The experimental scheme for evaluating the temperature considers heat conducted away through the specimen grips in two directions such that Q_T in Equation (9.2) can be divided into two components Q_{T1} and Q_{T2}. They correspond to the two ends of the specimen. According to the Fourier law [9], the amount of heat passing through an area dF over a period $d\tau$ is given by:

$$Q = \int_F \int_\tau -\lambda \frac{dT}{dX} \, dF \, d\tau \tag{9.3}$$

where λ is the heat conductivity coefficient, dT/dX the temperature gradient in the direction of heat flow. For a cylindrical specimen strained into the elastoplastic range while the grip portion remains in the elastic range, the temperature gradient is obtained from ΔT and ΔX which refers to points 2 and 4 about 5 mm apart as shown in Figure 9.1(a). This is made possible due to the linearity of the temperature distribution over the portion labelled 1 to 5 in Figure 9.1(a). Measurements were obtained for the period of increase deformation from τ_1 to τ_4 and decrease in deformation from τ_5 to τ_{10} when the specimen undergoes cooling. Note that linearity prevails for heating (solid lines) and cooling (dotted lines) as indicated in Figure 9.1(a). This gives Q_{T1} for the right end and Q_{T2} for the left end:

$$Q_{T1} = -\frac{\lambda F_1}{\Delta X_1} \int_0^\tau \Delta T_1 \, d\tau, \quad Q_{T2} = -\frac{\lambda F_2}{\Delta X_2} \int_0^\tau \Delta T_2 \, d\tau. \tag{9.4}$$

The total is

$$Q_T = Q_{T1} + Q_{T2}. \tag{9.5}$$

Thermocouples are welded to the remaining portion of the specimen to measure the temperatures T_1 to T_{10} as shown in Figure 9.1(b). Differential thermocouples are used at (T_1, T_2) and (T_3, T_4) such that ΔT_1 and ΔT_2 can be continuously recorded and used in Equation (9.4). The steady state condition corresponds to

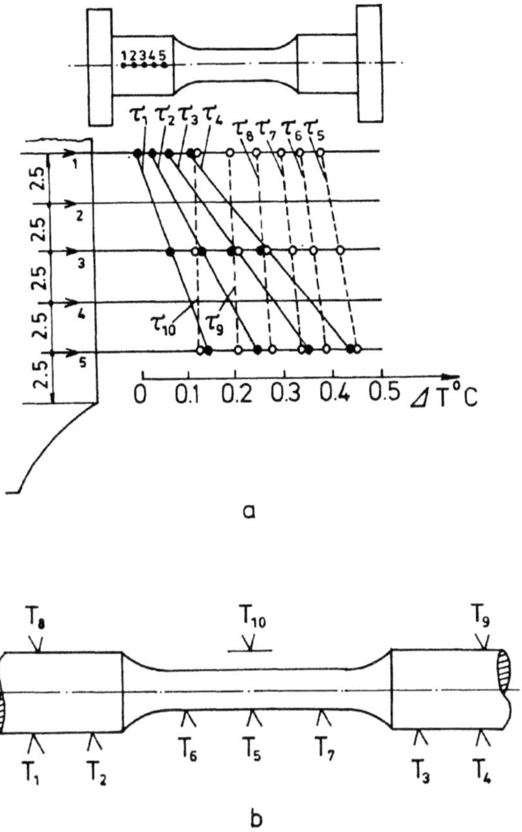

Fig. 9.1: Scheme of temperature measurement: (a) under loading (solid lines) and under cooling (dotted lines); (b) positions of thermocouple.

$\Delta T_1 = \Delta T_2 = 0$ which determines the limit of integration for Equation (9.4).

Referring to Figure 9.1(b), T_5 is the temperature at the specimen center while T_6 and T_7 are at an equal distance away from the center. The measured values of T_5 as a function of time is given in Figure 9.2(a). Time histories of $T_1 - T_2$ for static load and cyclic load are shown, respectively in Figure 9.2(b) and 9.2(c).

The heat generated within the specimen before it is conducted to its ends is given by [9]:

$$Q_P = C_p \cdot V \cdot \rho \cdot \Delta T \tag{9.6}$$

where ΔT is the temperature difference of the volume V before and after loading, C_p the specific heat capacity and ρ the density. When the steady state is reached, the thermocouples at T_5, T_6 and T_7 in the middle portion and those at the ends T_8 and T_9 register practically the same temperature. Note that T_8 and T_9 are located at the opposite sides of (T_1, T_2) and (T_3, T_4) where the differential thermocouples are

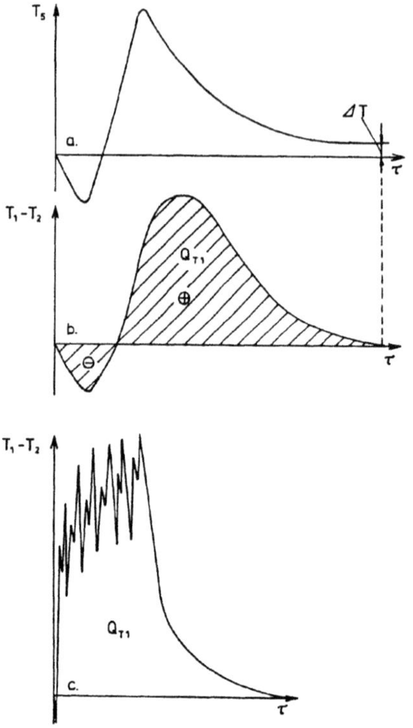

Fig. 9.2: Time history of temperature: (a) specimen center; (b) static load and (c) cyclic load.

situated. The thermocouple T_{10} in Figure 9.1(b) gives the temperature of the screen that is located at a distance away from the specimen; the temperature difference is negligibly small as mentioned earlier.

9.4 Static and cyclic loading

The temperature profiles for static and cyclic loading are quite different as shown in Figures 9.2(b) and 9.2(c). Thermal fluctuations in the case of cyclic loading were observed for a period of time. The smooth decay in $T_1 - T_2$ corresponds to the end of cyclic loading.

The foregoing procedure can be used to evaluate the elastoplastic deformation of specimens subjected to static and cyclic load. Specimens were made of 12X2MFA and X18H10T steel. The modified test device YMЭ−10T [10] equipped with a vacuum camera [11] were used. The deformation is recorded by extensometers [10] and photographing the loaded specimens at different time intervals.

Static load

Data on the uniaxial tension specimens are obtained up to fracture and they are given in Figure 9.3(a). Temperature measurements are also made for each step of loading by the method in [8]. Figure 9.3(b) gives the measured results for the total mechanical energy $\Sigma\overline{A}$, the energy dissipated in the form of heat $\Sigma\overline{Q}$ and absorbed energy $\Sigma\overline{E}$ in terms of the total strained volume. The quantities $\Sigma\overline{A}$, $\Sigma\overline{Q}$ and their difference $\Sigma\overline{E}$ tend to increase with increasing strain on the specimen; this dependence is approximately linear. The ratio $\Sigma\overline{Q}/\Sigma\overline{A}$ is almost constant during the deformation process. Figure 9.3(c) gives the value of 0.5 for the 12X2MFA steel and 0.35 for the X18H10T steel.

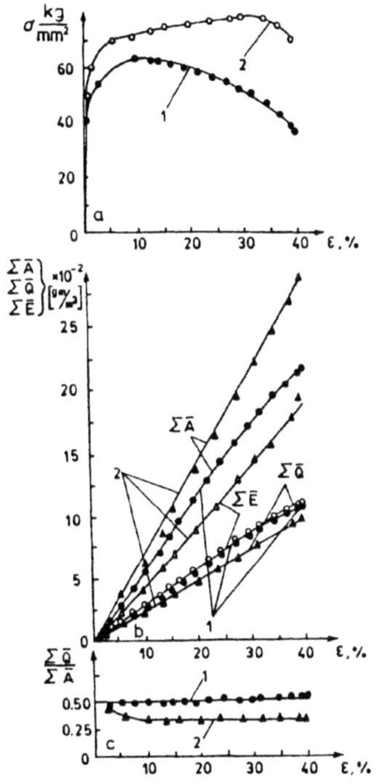

Fig. 9.3: Experimental data: (a) stress and strain; (b) specific energy components and (c) static fracture results for 12X2MFA (curves 1) and X18H10T steel (curves 2).

Cyclic load

Fatigue data for the 12X2MFA steel specimen subjected to a cyclic tension-compression load with a stress amplitude of $\sigma_a = 46$ kg/mm^2 are obtained; it fractured at $N = 137$ cycles. The width of the hysteresis loop δ is plotted in Figure 9.4(a) as a function of the number of cycles. It remained constant at first and then rose rapidly which means that the hysteresis loop area also increased. This is because of the increase of $\overline{A}_N$ for each subsequent segment of cyclic loads as seen in Figure 9.4(b). This character applies to $\overline{Q}_N$ and $\overline{E}_N$ being the difference of $\overline{A}_N$ and $\overline{Q}_N$ and their sums $\Sigma\overline{A}$, $\Sigma\overline{Q}$ and $\Sigma\overline{E}$ displayed in Figure 9.4(c). Similar results have been obtained for the X18H10T steel.

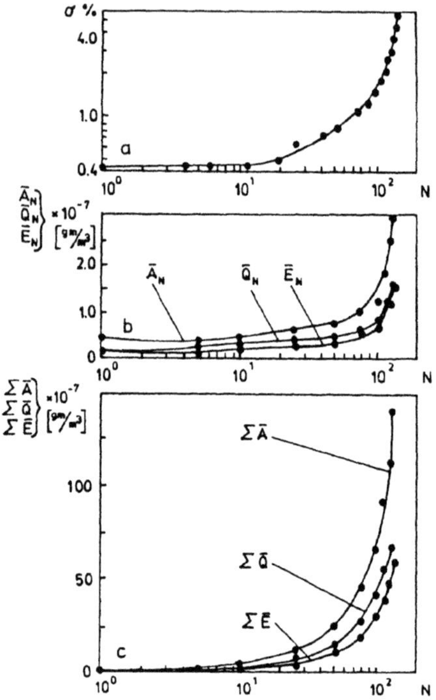

Fig. 9.4: Fatigue data for 12X2MFA steel: (a) cyclic plastic deformation; (b) specific energy components at each cycle and (c) accumulated values.

9.5 Local mechanical energy

Thus far, considerations have been given to energy averaged over the entire volume of the specimen, while fracture is a local phenomenon. Energy due to the deformation in the unfractured portion of the specimen does not contribute to fracture. Because energy components, when averaged over different portions of the specimen will not be the same, the determination of fracture energy should refer to the local volume near necking of the specimen where the actual fracture occurs. The evolution of deformation for the 12X2MFA steel specimen is shown in Figure 9.5(a) where the localized strain associated with neckiing leads to final fracture. Numerical values of the localized strain for the different stages of necking are shown in Figure 9.5(b); the strain at the neck can be seven (7) times larger than the mean value. This difference accounts for the difference in the specific mechanical energy based on V for the total volume of the strained material and V_ω at the necked region with a corresponding specific mechanical work

$$\overline{A}_\omega = \frac{1}{V_\omega} \int_{l_{0\omega}}^{l_{k\omega}} P \, dl \tag{9.7}$$

where $l_{0\omega}$ and $l_{k\omega}$ are the initial and the ultimate length of a local area where the neck forms and the fracture takes place. According to the experimental data in Figure 9.5(b), as the region of homogeneous strain may be taken approximately as $l_{0\omega} \simeq 0.1$ mm and $\overline{A}^e_\omega$ will denote the area in the $P = f(\varepsilon_\omega)$ plot shown in Figure 9.6(a) with the coordinate $P - \Delta l_\omega$. The corresponding values of A for 12X2MFA steel specimens of 8 mm diameter by 12 mm length and X18H10T tubular steel specimen of 21 mm diameter by 18 mm length are given in Table 9.1 which contain the values of $\Sigma\overline{A}$ for the total and specific mechanical energy of the entire strained volume used to cause static fracture.

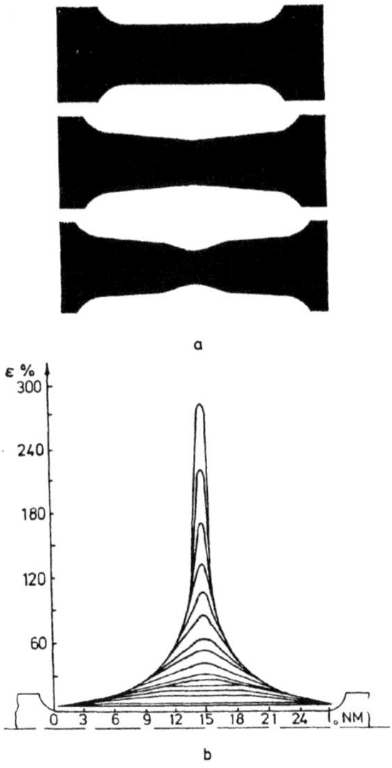

Fig. 9.5: Strains in static tension: (a) evolution of necking and (b) local strains.

Table 9.1: Specific mechanical energy.

Steel	A_Σ	$\overline{A}_\Sigma$ $\times 10^{-7}$	$\overline{A}^e_\omega$ $\times 10^{-7}$	$\overline{A}^p_\omega$ $\times 10^{-7}$	V $\times 10^{-6}$
	j	j/m^3	j/m^3	j/m^3	m^3
12X2MFA	--	--	151.7	135.4	1.76
	247.5	14.1	148.0	137.6	1.76
	245.4	13.9	157.7	141.4	1.76
	636.4	18.7	150.9	152.7	3.40
	670.0	21.9	129.0	128.2	3.05
X18H10T	1506.4	28.7	59.1	61.0	5.24

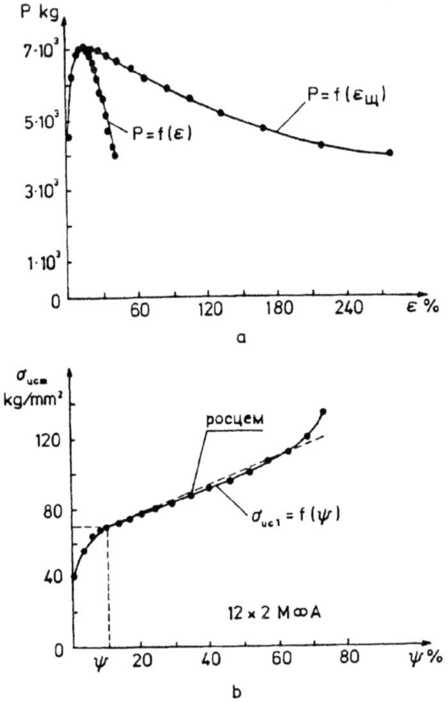

Fig. 9.6: The diagrams of specimen static tension with no total strain at its base; (a) strain at the neck zone and (b) true stresses versus transverse necking for 12X2MFA steel.

9.6 Analytical method

The analytical method in [12] may also be used to estimate the specific mechanical energy used to fracture the necked zone. It assumes that the energy used in static fracture can be obtained from the area 0 to σ_b. It can be regarded as one portion from 0 to σ_p and another one from σ_p to σ_b, i.e.,

$$\overline{A}_I^p = \delta_b \left[\frac{\sigma_p + 2\sigma_b}{3} \right]. \tag{9.8}$$

The specific work at a subsequent stage $\overline{A}_{II}^p$ until fracture may be estimated from the actual stress at necking where σ_b would be proportional to the relative change of the transverse cross-sectional area, $\Psi = 1 - F/F_0$. This is confirmed by the data for the 12X2MFA data in Figure 9.6(b) in which the solid curve corresponds to experiments and the dotted line may be used in design. The value of $\overline{A}_{II}^p$ is given by:

$$\overline{A}_{II}^p = \int_{\delta_b}^{\delta_k} \left[\frac{2\sigma_b(1 + \delta_b)}{1 + \delta_k} - \frac{\sigma_b(1 + \delta_b)^2}{(1 + \delta_k)^2} \right] d\delta. \tag{9.9}$$

The total work of deformation can be obtained by adding Equations (9.8) and (9.9)

$$\overline{A}_\omega^p = \overline{A}_I^p + \overline{A}_{II}^p = \frac{\delta_b}{3}(\sigma_p + 2\sigma_b) + 4.6\,\sigma_b(1 + \delta_b) \log \frac{1 + \delta_k}{1 + \delta_b} +$$

$$+ \sigma_b(1 + \delta_b)^2 \left[\frac{1}{1 + \delta_k} - \frac{1}{1 + \delta_b} \right]. \tag{9.10}$$

Table 9.1 shows that $\overline{A}^p$ computed from Equation (9.10) are in satisfactory agreement with experimental values $\overline{A}^e$ for both steels.

9.7 Local thermal energy

Determination of the specific thermal energy given off by fracture zone involves considering the nonhomogeneous character of the deformation. It will be associated with the local volume and strain. As pointed out in [8], the temperature elevation is nearly linear for increasing plastic strain. This dependence is shown in Figure 9.7 for the static tension of the two steels where the strains are measured by a transverse extensometer and evaluated subsequently in terms of the true longitudinal strain [13].

The decrease in temperature corresponds to the initial elastic deformation while heating corresponds to increase in the plastic strain. The relation

$$\Delta T = K \cdot e_{pl} \tag{9.11}$$

gives K a value of 1.12×10^2 grad.m/m for the 12X2MFA steel and to 0.82×10^2 grad.m/m for the X18H10T steel taking into account the small temperature drop for elastic strain [8]. The thermal energy for heating the material up to ΔT is given by $Q = C_p \rho V \Delta T$ and the corresponding strain is given by Equation (9.11). For the ith volume element V_i with homogeneous strain e_i

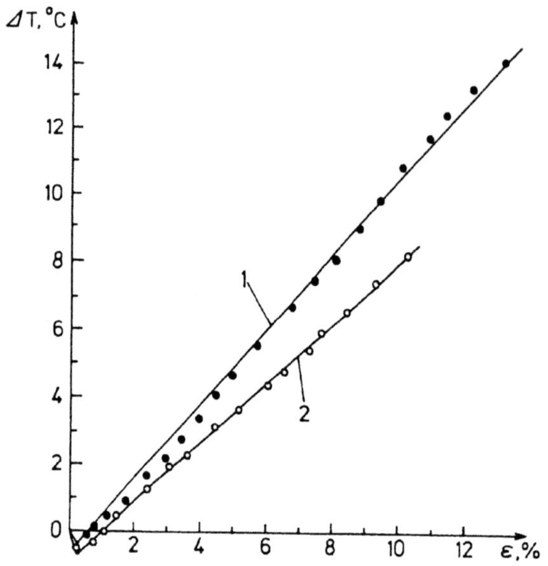

Fig. 9.7: Temperature versus strain for 12X2MFA (curve 1) and X18H10T (curve 2) steel.

Table 9.2: Measured data of Equation (9.14).

N Stage	Steel	1	2	3	4	5	6
$\eta^e = \frac{\sum_{i=1}^{n} v_i e_i}{\sum_{i=1}^{n} Q_i}$	12X2MFA	3.85	2.88	2.69	2.60	2.49	2.81
	X18H10T	4.09	4.47	3.73	3.43	3.87	2.50
N stage	Steel	7	8	9	10	11	12
$\eta^e = \frac{\sum_{i=1}^{n} v_i e_i}{\sum_{i=1}^{n} Q_i}$	12X2MFA	2.86	1.85	2.46	2.38	1.92	3.35
	X18H10T	3.40	2.74	3.05	2.66	2.65	1.20
N Stage	Steel	13	14	15	16	17	(mean value)
$\eta^e = \frac{\sum_{i=1}^{n} v_i e_i}{\sum_{i=1}^{n} Q_i}$	12X2MFA	1.33	2.33	1.19	1.83	0.86	2.33
	X18H10T	--	--	--	--	--	3.15

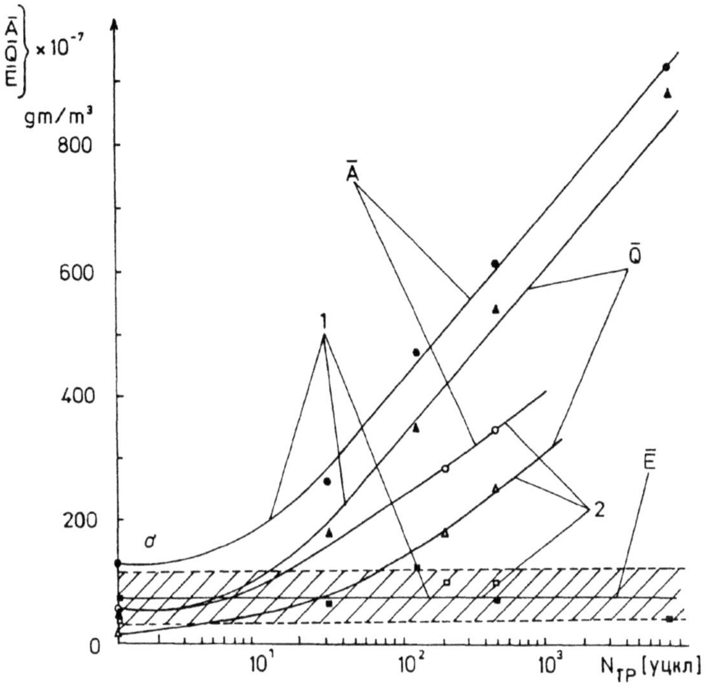

Fig. 9.8: Variations of specific energy components with number of cycles for 12X2MFA (curves 1) and X18H10T (curves 2) steel.

$$Q_i = C_p \cdot \rho \cdot V_i \cdot K \cdot e_i. \tag{9.12}$$

Summation of the volumes for the entire specimen yields

$$\sum_{i=1}^{n} Q_i = C_p \cdot \rho K \sum_{i=1}^{n} V_i \cdot e_i \tag{9.13}$$

which is the total thermal energy. The ratio

$$\frac{\sum_{i=1}^{n} V_i e_i}{\sum_{i=1}^{n} Q_i} = \frac{1}{C_p \cdot \rho \cdot K} = \eta \tag{9.14}$$

can be defined for a specific material; it equals 2.48×10^{-9} m$^3/j$ for the 12X2MFA steel and to 3.28×10^{-9} m$^3/j$ for the X18H10T steel. Table 9.2 gives some additional data on static fracture where η^e are measured for strains in small areas and η_{cp} are mean values; they are both nearly equal to η. This validates the estimate on thermal energy and the corresponding strain. The specific thermal energy at fracture of the necked zone can be expressed in terms of the true local logarithmic strain by using Equation (9.12). Using the data on the specific mechanical energy at fracture $\overline{A}_\omega$ already known and $\overline{Q}_\omega$, Equation (9.1) can be applied to obtain $\overline{E}_\omega$ which is 75.3×10^7 g/m^3 for the 12X2MFA steel and 19.1×10^7 g/m^3 for the X18H10T steel.

More difficulties would be encountered for the case of cyclic loading because strain redistribution takes place repeatedly before final fracture; the situation is not completely static. Kinetics will affect the fracture energy. If the specimen is sufficiently stiff where necking does not occur between crack initiation and propagation, then a homogeneous state of strain may be assumed. The sum of all areas under the load deformation diagrams for all cycles will give the specific mechanical energy $\overline{A}$ while the specific thermal energy $\overline{Q}$ can be obtained from known values of the cyclic plastic strain e_{pl} and K according to Equation (9.12). The difference between $\overline{A}$ and $\overline{Q}$ gives $\overline{E}$. Figure 9.8 gives estimates of $\overline{A}$, $\overline{Q}$ and $\overline{E}$ for the tested materials having different lives. It is seen that for both materials, the specific energies increase with N while they remain constant under static fracture with some scatter. The value $\overline{A}_p = 101.5 \times 10^7$ g/m^3 may be taken as the specific fracture energy for iron [14]. A criterion for characterizing material life has thus been presented.

References

[1] ROMANOV, A.N., Energy of fracture for cyclic loading, *Problemy protchnosti,* No. 3, 1971, pp. 3–9 (in Russian).

[2] ROMANOV, A.N., Energy criteria of fracture for low-cyclic loading, *Problemy protchnosti,* No. 1, 1974, pp. 3–18 (in Russian).

[3] FASTOV, N.S., Heat conduction in plastic deformation, *Doklady AN SSSR (novaj seria),* Vol. XXXIII, No. 6, 1952, pp. 851–854 (in Russian).

[4] GRIFFITH, A.A., The phenomena of rupture and flow in solids, *Philosophical Transactions Royal Society,* Ser. A, Vol. 221, 1920–21, pp. 163–168.

[5] OROWAN, E., Energy criteria of fracture. Modifications of the Griffith theory are presented to cover the case for a rapidly running crack and for starting in a stationary crack, *Welding Journal,* No. 3, 1955, pp. 157–160.

[6] RICE, J.R., Path independent integral and approximate analysis of strain concentration by notches and cracks, *ASME Paper* 68–APM–31, 1968, p. 8.

[7] FEDOROV, V.V., Thermodynamic of strength and fracture of solid copper, *Problemy protchnosti,* No. 11, 1971 pp. 32–34 (in Russian).

[8] GADENIN, M.M. and ROMANOV, A.N., Method for experimental determination of energy in static and cyclic elasto-plastic deformation and fracture, *Zavodskaj laboratoria,* No. 8, 1978, pp. 997–1002 (in Russian).

[9] ROMANOV, A.N., GADENIN, M.M. and JUNIN, V.M., High-temperature vacuum programmable machines for investigation of low-cyclic fracture processes. *Temperature Microscopic Investigation of Metals and Alloys,* Moscow, Nauka, 1974, pp. 13–19 (in Russian).

[10] ROMANOV, A.N., High-temperature vacuum camera for low-cycle testing machine under tension-compression, *Zavodskaj laboratoria,* Vol. XXXVII, No. 6, 1971, pp. 727–730 (in Russian).

[11] JILIMO, L., Construction steel properties. Characterizing by work of limit strain, *New Problems of Metallurgy,* Moscow, An SSSR, 1958, pp. 572–582 (in Russian).

[12] GADENIN, M.M. and ROMANOV, A.N., *Range of Strain under Tension-Compression Cyclic Elasto-Plastic Deformation. Structure Factors under Low-Cycle Fracture of Metals,* Moscow, Nauka, 1977, pp. 115–129 (in Russian).

[13] IVANOVA, V.S. and TERENTIEV, V.F., *Nature of Metal Fatigue,* Moscow, Metallurgia, 1975 (in Russian).
[14] JASTRGEMBSKI, A.S., *Technical Thermodynamics,* Moscow-Leningrad, Gosenergoizdat, 1953 (in Russian).

V.D. Klushnikov

10

Stability of materials and structures

10.1 Introduction

'Stability' in solid mechanics refers to the behavior of a solid to maintain certain configuration under the action of a load. A change in the solid configuration and/or morphology, therefore, corresponds to an instability that is caused by sustained or increased loading. The onset of instability defines the load carrying capacity or integrity of a structure. Material instability caused by deformation can also occur regardless of the loading type. In what follows, structural instability will be distinguished from that of material instability.

Internal material instability can be related to irregularities or defects that can lead to fracture in materials such as rocks. Damage can be path dependent and can affect the mechanical properties throughout the loading history. As continuity of the material can be interrupted, the ways with which experimental data are collected and used in continuum mechanics theories should be exercised with care as the physical process may violate some of the initial assumptions.

10.2 Stability of elastic-plastic structures

A criterion of stability is associated with changes in structure configuration due to the action of load. For elastic structures, instability is related to an alteration in the equilibrium state, such as buckling, while instability of plastic structures results from the deformation process [1]. The criteria of stability, therefore, differ. Elasticity concerns bifurcation of the state and plasticity is related to bifurcation of the process. The former involves determining the zero order of nonunique state and the latter involves finding the first-order nonunique increments of change of state parameters of the state.

Nonuniqueness of a solution due to bifurcation can be determined by considering infinitely small disturbances. The linearized equations of statics for a continuum may be used:

$$\operatorname{div}(\sigma + A\sigma^0\delta g) + F = 0, \quad \delta g = \operatorname{grad}\delta u,$$

$$(\sigma + B\sigma^0\delta g)\nu^0\Big|_{S_T} = T, \quad \delta u\Big|_{S_u} = \delta U \tag{10.1}$$

where σ and σ^0 are the stress tensors referred to the current and initial state, respectively, while F is the body force vector and T the traction vector. The variation of the displacement vector from the initial state in the body is given by δu and on the surface by δU. In Equations (10.1), A and B are fourth order tensors whose components are constants. The surface of the body is $S = S_T + S_u$ with the unit normal vector ν such that T is specified on S_T and U on S_u.

Basic approach

For the initial state $F = F^0$, $T = T^0$ and $\delta u = 0$, the system in Equation (10.1) must therefore satisfy $\sigma = \sigma^0$ and $\delta u = 0$. For an infinitesimal increment $\delta\sigma$, it follows that

$$\operatorname{div}(\delta\sigma + A\sigma^0\delta g) + \delta F = 0$$

$$(\delta\sigma + B\sigma^0\delta g)\nu\Big|_{S_T} = \delta T, \quad \delta u\Big|_{S_u} = \delta U. \tag{10.2}$$

Making use of the constitutive equation, the stress increments $\delta\sigma = \sigma - \sigma^0$ can be related to the strain increments $\delta e = 1/2\delta(g + g^{T})$. Using Equations (10.2), the problem of the bifurcation of the zero order may be solved. In the case when the external forces F and T do not vary, then $\delta T = T - T^0$, $\delta F = F - F^0 = 0$ and $\delta u = 0$ in Equations (10.2). In general, slight changes do prevail in $\delta\sigma$, δe and δu. These changes, however, are not sufficient for determining the complete bifurcation problem; other methods may have to be used.

Bifurcation process of the first order consists of finding the nonuniqueness of internal state parameters by considering the nonzero variations δF^0, δT^0 such that

$$\Delta\sigma = \delta\sigma - \delta\sigma^0, \quad \Delta e = \delta e - \delta e^0, \quad \Delta u = \delta u - \delta u^0. \tag{10.3}$$

Applying Equations (10.2) and (10.3), these results

$$\operatorname{div}(\Delta\sigma + A\sigma^0\Delta g) = 0, \quad \Delta g = \operatorname{grad}\Delta u,$$

$$(\Delta\sigma + B\sigma^0\Delta g)\nu\Big|_{S_T} = 0, \quad \Delta u\Big|_{S_u} = 0. \tag{10.4}$$

The symbol Δ refers to infinitesimal incremental change. For an elastic-plastic solid, the $\Delta\sigma$ versus Δe relation cannot be stated uniquely. As in the case of bifurcation, a precise relation between $\Delta\sigma$ and Δe requires a knowledge of the plastic zones associated with the current configuration of the body. Depending on the smoothness of the solution, the relation $\Delta\sigma \sim \Delta e$ can degenerate to $\delta\sigma \sim \delta e$ at discrete points [1, 2]. Such points were first found experimentally in [3] and independently by other means [4]. These findings establish the rule for calculating the critical loads for elastic-plastic structures and the results agree well with the experiments.

A theoretical justification of the above approach required additional work [5, 6]. Disturbed motions of elastic-plastic systems under continuous loading were found later in [3, 7]. As it turns out, it is possible to have discrete points of instability within finite time intervals which tends to indicate that continuous dependence of motion on the initial condition [1] is not completely satisfied, i.e., at a given instant, the motion can be unstable. Vanishingly small disturbances can thus be superimposed on the motion at a given instance giving rise to finite deviations for finite time intervals. This applies at least for simple structures where a curved rod can become straight.

Higher order bifurcation

While the state parameters for the first order bifurcation, denoted as $B1$, may be unique, the formulation may be expressed in terms of nonuniqueness of the velocity components instead of the displacements. It follows that the second order bifurcation, $B2$, refers to nonuniqueness of the acceleration rates. Conceivably, bifurcation of the Mth order [1], i.e., BM may lead to possible new findings in the deformation process.

If the medium is elastic, only $B0$ prevails. The $B0$ points do not coincide with the points $B1$ for an elastic-plastic medium while the points BM for $M \geq 2$ coincide with those of $B1$. For high order differential constitutive equations, it is impossible to have bifurcation points limited below a certain order.

Frequently, it may be more expedient to evaluate higher order bifurcation points. For instance, the stability of the elastic state [8] may be determined from $B1$ even though the problem is $B0$. In other situations, the nonlinear plasticity problem associated with $B1$ may be identified with $B2$ that involves acceleration rates of the constitutive relation becomes linear. This is true, in general, because plasticity deals with strain rates. Hence, stability involving plasticity can be solved in terms of $B2$ while the post-critical behavior may be $B1$.

Let the constitutive equation of plasticity be expressed in tensorial form:

$$f(\sigma, \dot{\sigma}, e, \dot{e}) = 0. \tag{10.5}$$

The $B2$ problem ($\Delta\sigma = \Delta e = \Delta\dot{\sigma} = \Delta\dot{e} = 0$) refers to the plastic domain

$$f_{\dot{\sigma}} \Delta\ddot{\sigma} + f_{\dot{e}} \Delta\ddot{e} = 0 \tag{10.6}$$

where

$$\Delta\ddot{\sigma} = \ddot{\sigma} - \ddot{\sigma}^0, \quad \Delta\ddot{e} = \ddot{e} - \ddot{e}^0, \quad \left(f_a = \frac{\partial f}{\partial a}\right). \tag{10.7}$$

In elasticity, $\Delta\ddot{\sigma} = C\Delta\ddot{e}$ where C is a fourth order tensor whose components are constants. Equation (10.6) can be expressed in the form

$$\Delta\ddot{\sigma} = \tilde{C}\Delta\ddot{e}, \quad \tilde{C} = \begin{cases} C, & (E_0) \\ C', & (P_0) \end{cases} \tag{10.8}$$

provided that $\tilde{C}$ is now a fourth order tensor that depends on σ^0, e^0; the rates $\dot{\sigma}^0$, $\dot{e}^0$, at a given instance; and the configurations of the domains E_0 and P_0 are given. If Equation (10.5) is linear in $\dot{\sigma}$ and $\dot{e}$, then $\tilde{c}$ is independent of rates and the $B1$ problem involves

$$\Delta\dot{\sigma} = \tilde{C}\Delta\dot{e}. \tag{10.9}$$

The form of Equation (10.4) applies also the BM problem; the quantities $\Delta\sigma$ and Δu are replaced respectively by $\Delta\sigma^{(M)}$ and $\Delta u^{(M)}$. For example, the form

$$\operatorname{div}(\Delta\dot{\sigma} + A\sigma^0\Delta\dot{g}) + \operatorname{div}(A\overset{\cdot}{\sigma}{}^0\Delta g) = 0 \qquad (10.10)$$

is preserved for $B1$ with $\Delta g = 0$. Hence, if relations expressed by Equation (10.8) for the problem BM can be found, then plasticity coincides with the problem $B0$ for nonhomogeneous elastic solids. In this sense, it appears that there is an elastic equivalent to the stability of plastic solids. This possibility, however, does not exist even though the functional form of Equation (10.8) so suggests because E_0 and P_0 cannot be preserved so that nonlinearity comes into play.

10.3 Stability of visco-elastic structures

If the tensorial function f in Equation (10.5) is not homogeneous in $\dot{\sigma}$ and $\dot{e}$, which is a boundedness requirement in plasticity, then Equation (10.5) refers to the constitutive equation for some visco-elastic solids. In this case together with the condition

$$\Delta\sigma \ll \sigma^0, \quad \Delta e \ll e^0, \quad \Delta u \ll u^0. \qquad (10.11)$$

We can demand that

$$\Delta\dot{\sigma} \ll \dot{\sigma}^0, \quad \Delta\dot{e} \ll \dot{e}^0, \quad \Delta\dot{u} \ll \dot{u}^0. \qquad (10.12)$$

Even though Equation (10.5) is nonlinear in the rates, then $B1$ problem turns out to be linear because of the condition

$$f_\sigma\Delta\sigma + f_{\dot{\sigma}}\Delta\dot{\sigma} + f_e\Delta e + f_{\dot{e}}\Delta\dot{e} = 0. \qquad (10.13)$$

It can be seen for the case of $B1$ that Equations (10.12) are not satisfied. The same applies to $B2$. For visco-elastic structures under creep, the stability process can be interpreted in the Ljapunov sense [9]. Critical levels of applied forces can be found for visco-plastic solids. The results, however, do not depend continuously on the initial data if the process is extended to infinity. In practice, it is of interest to know the behavior of the structures for finite time interval. For an elastic-plastic solid, it is possible to disturb the continual dependence of the result on initial conditions for finite time interval. This occurs at infinity if the solid is visco-elastic.

Referring to Equations (10.5), (10.11), (10.12) and σ^0, it is possible to identify with the conditions

$$\Delta\dot{\sigma} = \Delta\dot{e} = \Delta\dot{u} = 0 \qquad (10.14)$$

in which $\Delta\sigma$, Δe and Δu are different from zero. If Equations (10.13) enforced by the conditions in Equation (10.14) can be solved in terms of $\Delta\sigma$, then

$$\Delta\sigma = \tilde{C}\Delta g. \qquad (10.15)$$

The matrix $\tilde{C}$ involves the derivatives f_σ and f_e while the points of instability governed by Equation (10.14) can be identified as the $B0$ problem for equivalent elastic case, Equation (10.5). The critical conditions [1] will contain, besides the external forces, the derivatives of f, displacements and other parameters associated with the deformation process. Even if the load is constant, the displacements can be time dependent and the time corresponding to the critical condition can be determined.

Equation (10.10) must also be satisfied in addition to Equation (10.4). No additional kinematic limitations prevail only if $\dot{\sigma}^0 = 0$.

Discrete instances of disturbed motion for the simplest visco-elastic structure can be found [1] in the same way as that in [11]. These are referred to as pseudobifurcation points (*PB*) because the disturbances are vanishingly small. By means of differentiation, a succession of *PB* points can be found. For the *PB* point of the first order denoted as *PB* 1, the conditions to be satisfied are

$$\Delta\ddot{\sigma} = \Delta\ddot{e} = \Delta\ddot{u} = \Delta\sigma = \Delta e = \Delta u = 0$$

such that $\Delta\dot{\sigma}$, $\Delta\dot{e}$ and $\Delta\dot{u}$ are different from zero. Assuming as before that $\dot{\sigma}^0 = 0$ for the relation $\Delta\dot{\sigma} \sim \Delta\dot{e}$, then Equation (10.5) yields

$$(f_\sigma + f_{\sigma e}\dot{e} + f_{\dot{\sigma}\dot{e}}\ddot{e})\Delta\dot{\sigma} + (f_e + f_{e\dot{e}} + f_{\dot{e}\dot{e}}\ddot{e})\Delta\dot{e} = 0$$

so that the elastically equivalent matrix for *PB* 1 and *PB* 0 are different and hence the points *PB* 1 and *PB* 2 are also different. An increase in the order of the constitutive equation for *PB* 1 does not yield any additional points. Analogous to *PB* 1 is the stability criterion for creep [12]. The transition of undisturbed to disturbed motion takes place slowly after which acceleration occurs. This procedure establishes the equivalent elastic definition of *PB* of high orders. A generalization of differential constitutive equation of any order different from Equation (10.5) can be made in addition to constitutive equations in integral form [1]. The order of the pseudobifurcation points increases with time. In general, pseudobifurcation is not as pronounced as bifurcation. It follows that the visco-elastic system associated with constant external forces can overlook several *PB* points; this occurs for higher order deformation. In contrast to bifurcation, the pseudobifurcation approach is identified by a whole set of critical points.

For small strains, the equivalent elastic representation is

$$\Delta e' = a\Delta S + b(s^0\Delta S)S^0 \qquad (10.16)$$

in which $\Delta e'$, $\Delta S'$ and S^0 are deviators of the respective strain and stresses, i.e., Δe, $\Delta\sigma$ and σ^0 while a, b and $S^0\Delta S = S^0_{mn}\Delta S_{mn}$ are scalars. For an incompressible material, $\Delta e' = \Delta e$, Equation (10.16) fully describes the system. The simplest deformation and incremental theory of plasticity with isotropic hardening correspond to

$$a = \frac{1}{2G_*},\quad f = \frac{1}{4T_0^2}\left(\frac{1}{G'},\ -\frac{1}{G_*}\right),\quad T_0^2 = \frac{1}{2}S'^0_{mn}S^0_{mn} \qquad (10.17)$$

where $G_* = G_s$ is the secant modulus in the deformation theory and $G_* = G$ is the elastic shear modulus in the incremental theory. In both cases, G' is the tangential shear modulus. Equation (10.16) gives the equivalent of *B* 1 which can be extended to *BM* ($M \geq 2$). The quantities relevant to creep [13] are

$$P^\alpha \dot{P}_{ij} - AT^{n-1}S_{ij},\quad P_{ij} = e'_{ij} - \frac{S_{ij}}{2G}$$

$$\dot{P} = \sqrt{2P_{ij}P_{ij}}, \quad T^2 = \frac{1}{2}S_{ij}S_{ij} \tag{10.18}$$

for BM $(M \geq 1)$ which also apply to $B\,0$ for the elastic solid. For PB of the Nth order, it follows that

$$a = \frac{1}{2}\left(\frac{1}{G} + \frac{P_0}{2\alpha T_0 N}\right), \quad b = \frac{P_0}{4\alpha T_0^3 N}\left(n - 1 - \frac{n}{1 + N}\right). \tag{10.19}$$

The following integral equation refers to the hereditary theory:

$$e_{ij} = \frac{S_{ij}}{2G} + \frac{1}{2G}\int_{t_0}^{t} K(t, \tau)S_{ij}(\tau)\,\mathrm{d}\tau. \tag{10.20}$$

For $S_{ij}(\tau)$ in the vicinity of $\tau = t$, $b = 0$ and

$$a = \frac{1}{2G}\left[1 + \frac{1}{N!}\int_{t_0}^{t} K_t^{(N)}(t, \tau)(\tau - t)^N\,\mathrm{d}\tau\right] \tag{10.21}$$

where $PB = \infty$ is the bifurcational point. For details, refer to [1].

10.4 Stability of a strip

More recently, different forms of Equations (10.4) have been used to investigate the stability of three-dimensional bodies. Among them are

$$A\sigma^0\Delta g \sim \overset{0}{\sigma}_{ik}\Delta g_{jk}, \quad B\sigma^0\Delta g \sim \begin{cases}\overset{0}{\sigma}_{ij}\Delta g_{jk}\\ 0\end{cases} \tag{10.22}$$

and those of Leubenzon–Ishlinsky [14]:

$$A\sigma^0\Delta g = 0, \quad B\sigma^0\Delta g \sim \begin{cases}-\overset{0}{\sigma}_{jk}\Delta g_{ik}\\ -\overset{0}{\sigma}_{jk}\Delta g_{ik} - \overset{0}{\sigma}_{ik}\Delta g_{kj}\end{cases}. \tag{10.23}$$

Referring to the definitions of the infinitesimal displacements and rotations given by

$$\Delta e = \frac{1}{2}(\Delta g + \Delta g^T), \quad \Delta\omega = \frac{1}{2}(\Delta g - \Delta g^T) \tag{10.24}$$

consider the approximation

$$\Delta g = \Delta e + \Delta\omega \cong \Delta\omega \tag{10.25}$$

such that

$$\Delta\omega = -\Delta\omega^T \tag{10.26}$$

then Equation (10.22) may be written as

$$A\sigma^0\Delta g \sim -\overset{0}{\sigma}_{ik}\Delta g_{kj}, \quad B\sigma^0\Delta g \sim \begin{cases}-\overset{0}{\sigma}_{ik}\Delta g_{kj}\\ 0\end{cases}. \tag{10.27}$$

If body forces in the first of Equations (10.4) are absent, then

$$(\Delta\sigma_{ij} - \sigma^0_{ik}\Delta u_{k,j})_{,i} = \Delta\sigma_{ij,i} - \sigma^0_{ik}\Delta u_{k,ij}$$
$$= \sigma_{ij,j} - \sigma^0_{ik}\Delta e_{ik,j}. \qquad (10.28)$$

It follows that $\Delta e \sim \Delta\sigma$ can be represented as in the case of linear elasticity and Equations (10.21) and (10.22) can be formulated in terms of the stresses.

Consider the case of homogenous deformation and incompressible material. Making use of Equation (10.6), it is found that

$$\sigma^0_{ik}\Delta e_{ik} = S^0_{ik}[a\Delta S_{ik} + bS_{ik}(S^0_{mn}\Delta_{mn})]$$
$$= (a + 2bT_0^2)(S^0_{mn}\Delta S_{mn}). \qquad (10.29)$$

in which

$$C = a + 2bT_0^2, \quad \phi = S^0_{mn}\Delta S_{mn},$$
$$\overset{*}{\sigma}_{ij} = \Delta\sigma_{ij} - C\phi\,\delta_{ij}, \quad \overset{*}{\sigma} = \Delta\sigma - c\phi, \quad (\Delta\sigma = \frac{1}{3}\Delta\sigma_{kk}). \qquad (10.30)$$

Equilibrium and boundary conditions are given by

$$\overset{*}{\sigma}_{ij,i} = 0, \quad \overset{*}{\sigma}_{ij}\nu_j + c\phi\,\nu_i = 0. \qquad (10.31)$$

The quantity $\overset{*}{\sigma}$ differs from $\Delta\sigma$ by the isotropic part such that $\Delta S_{ij} = \overset{*}{S}_{ij}$. Strain compatibility and Equation (10.16) yield

$$a(\overset{*}{S}_{ij,ks} - \overset{*}{S}_{jk,si} + S_{ks,ij} - S_{si,jk}) + b(S^0_{ij}\phi_{,ks} - S^0_{jk}\phi_{,si} +$$
$$+ S^0_{ks}\phi_{,ij} - S^0_{si}\phi_{,jk}) = 0. \qquad (10.32)$$

The first of Equation (10.31) gives

$$\overset{*}{S}_{jk,k} = -\overset{*}{\sigma}_{,j}, \qquad (10.33)$$

which, together with Equation (10.32) lead to

$$\overset{*}{S}_{ij,kk} + 2\overset{*}{\sigma}_{,ij} = \frac{b}{a}(S^0_{ik}\phi_{,kj} - S^0_{ij}\phi_{,kk} + S^0_{jk}\phi_{,ki})$$
$$\overset{*}{\sigma}_{,kk} = \frac{b}{a}S^0_{mn}\phi_{,mn}. \qquad (10.34)$$

The condition of plane strain corresponds to $\Delta e_{33} = \Delta S_{33} = 0$, $S^0_{11} + S^0_{22} = 0$. The first of Equation (10.31) can be satisfied by letting

$$\overset{*}{\sigma}_{11} = F_{,22}, \quad \overset{*}{\sigma}_{22} = F_{,11}, \quad \overset{*}{\sigma}_{12} = -F_{,12}. \qquad (10.35)$$

It can be shown that

$$\phi = S^0_{mn}\overset{*}{S}_{mn} = S^0_{mn}\overset{*}{\sigma}_{mn} = -S^0_{mn}F_{,mn}. \qquad (10.36)$$

The second of Equations (10.31) therefore gives

$$F_{,ijij} + \frac{b}{a}S^0_{ij}S_{mn}F_{,ijmn} = 0. \qquad (10.37)$$

A strip of length $2l$ and thickness $2h$ is stretched longitudinally in the x_1-direction. In this case, $S^0_{22} = -S^0_{11} = -S_0$, $T_0 = S_0$ and Equation (10.37) takes the form

$$\frac{\partial^4 F}{\partial x_1^4} + 2\kappa^2 \frac{\partial^4 F}{\partial x_1^2 \partial x_2^2} + \frac{\partial^4 F}{\partial x_2^4} = 0 \tag{10.38}$$

where

$$\kappa^2 = \frac{a - bS_0^2}{a + bS_0^2}. \tag{10.39}$$

Boundary conditions at $x_2 = \pm h$ and Equation (10.32) render

$$\frac{\partial^2 F}{\partial x_1 \partial x_2} = 0, \quad \frac{\partial^2 F}{\partial x_1^2} - CS_0\left(\frac{\partial^2 F}{\partial x_1^2} - \frac{\partial^2 F}{\partial x_2^2}\right) \tag{10.40}$$

whose solution takes the form

$$F = (A \cos \gamma z + A \cos \overline{\gamma} z) \cos \lambda x_1 \tag{10.41}$$

where

$$z = Ax_2, \quad \gamma = \alpha + i\beta, \quad i = \sqrt{-1}$$

$$\alpha^2 = \frac{1 - \kappa^2}{2}, \quad \beta^2 = \frac{1 + \kappa^2}{2}. \tag{10.42}$$

The first of Equations (10.40) is satisfied if

$$A = \frac{B}{\gamma} \sin \overline{\gamma} \lambda h, \quad \overline{A} = -\frac{B}{\overline{\gamma}} \sin \gamma \lambda h. \tag{10.43}$$

The second of Equations (10.40) gives

$$\delta_0 = \frac{1}{2C}\left(1 + \frac{\alpha \sinh 2\beta\lambda h}{\beta \sin 2\alpha\lambda h}\right). \tag{10.44}$$

Boundary conditions at the ends of the strip $x_1 = \pm l$ are satisfied by taking $\lambda l = \pi/2$. For a sufficiently long strip, the second term in parentheses in Equation (10.44) is small compared to unity and hence

$$S_0 \cong \frac{1}{C} = \frac{1}{a + bS_0^2}. \tag{10.45}$$

In view of Equation (10.17) for the simple case, $S_0 = 2G'$ prevails. Within the bounds of the creep theory in Equation (10.18) with $S_0 = \text{const.}$, Equation (10.45) gives the critical value P_0. It is found from Equations (10.19) and (10.45) with $T_0 = S_0$ that

$$P_0 = \frac{2\alpha(1 + N)}{n}\left(1 - \frac{S_0}{2G}\right). \tag{10.46}$$

Even if $S_0/2G$ is small in comparison with unity, strain cannot be considered to be small for α/h is larger than 0.1.

According to the hereditary theory, Equations (10.21) and (10.45) give

$$\int_{t_0}^{t} K_t^{(N)}(t, \tau)(\tau - t)^N \, d\tau = N!\left(\frac{2G}{S_0} - 1\right). \tag{10.47}$$

Equation (10.23) leads to the same result (see [15]). Equation (10.22) disallows necking. Buckling, together with thickness change of the strip, can be found. The critical condition can be obtained from Equation (10.44) for small h/l, i.e.,

$$S_0 = -\frac{\pi^2}{12C}\left(\frac{h}{l}\right)^2. \tag{10.48}$$

The same critical condition can be found from the theory of plates. If strain state is not plane for stretching in two directions, Equation (10.34) can be expressed in terms of the function

$$\phi_{,ijij} + \frac{2b}{C} S_{ij}^0 (S_{mn}^0 \phi_{,ijmn} - S_{ik}^0 \phi_{,kjmn}) = 0. \tag{10.49}$$

The components $\overset{*}{\sigma}_{ij}$ can be defined by using Equation (10.34) and the appropriate boundary conditions.

10.5 Macroscopic material stability

Material instability differs from that of structural instability; it has been considered in [16]. Instability refers to the loss of load carrying capacity and certain critical condition of the stress and strain state. The constitutive equation of the material must consider the effects of dilatation and internal friction for determining the structural instability of the material. A stress and strain diagram alone may not contain sufficient information. Violation of continuity [17] and stability [18] must be considered while macroexperiments must also be performed, but these are not able to resolve the issue completely. This is because the condition of continuity in the incremental theory [19] can give rise to singularities [20] for surface loadings. This was studied in [21] by equivalent loadings and considering constitutive equation in hypoelasticity [22].

In addition to the discrete nature of loading, the history of the deformation process must also be considered. For an elastic-plastic solid, the rates of plastic strain and strain rates are related as

$$\dot{e}_{ij}^{p} = \frac{\partial W(\sigma_{mn}, \dot{\sigma}_{mn}, \chi)}{\partial \dot{\sigma}_{ij}}. \tag{10.50}$$

The parameter χ defines the loading history [23]. If $f(\sigma_{mn}, e_{mn}^{p}, \chi) = 0$ corresponds to the case of smooth loading surface, then continuity requires

$$\dot{e}_{ij}^{p} = Q_{ij}\frac{\partial f}{\partial \sigma_{mn}}\dot{\sigma}_{mn}. \tag{10.51}$$

Substituting this relation into the condition

$$\frac{\partial \dot{e}_{ij}^{p}}{\partial \dot{\sigma}_{mn}} = \frac{\partial \dot{e}_{mn}^{p}}{\partial \dot{\sigma}_{ij}} \tag{10.52}$$

by using Equation (10.50), it is found that

$$A_{ijmn} = Q_{ij}\frac{\partial f}{\partial \sigma_{mn}} = Q_{mn}\frac{\partial f}{\partial \sigma_{ij}} = A_{mnij}. \tag{10.53}$$

This leads to the result

$$Q_{ij} = q\frac{\partial f}{\partial \sigma_{ij}}. \tag{10.54}$$

Equation (10.51) shows that $A_{ijmn} \neq A_{mnij}$. Let a prime denote the antisymmetric part so that W and W' can be written as

$$W = \int G_{ij} e_{ij}^{p}\,\mathrm{d}\,t,$$

$$W' = \int \sigma_{ij} A'_{ijmn} \dot{\sigma}_{mn}\,\mathrm{d}\,t = \frac{1}{2}\int A_{ijmn}(\sigma_{ij}\mathrm{d}\sigma_{mn} - \sigma_{mn}\,\mathrm{d}\sigma_{ij}).$$

When $\sigma_{ij} = m\sigma_{ij}^{0}$, the increment W' is equal to zero. This also occurs if $(\partial f/\partial \sigma_{ij})$ $(\partial f/\partial \sigma_{ij})\,\mathrm{d}\,\sigma_{ij} = 0$.

Equations (10.51) and (10.54) correspond to the certain regularity of the loading surface. Change of the loading surface must be continuous such that

$$\dot{f} = f_{\sigma}\dot{\sigma} + f_{e}\dot{e}^{P} + f_{\chi}\dot{\chi} = 0$$

which defines q and f in Equation (10.54). Additional considerations are required if small change in strain cause only small change in the loading surface with singularities. Accumulated effects may be considered [24]. Material instability can also be studied without using the concept of loading surface [25, 26].

References

[1] KLUSHNIKOV, V.D., *The Stability of Elasto-Plastic Systems,* Nauka, Moskva, 1980, p. 240.

[2] KLUSHNIKOV, V.D., *Bifurcation of the Process of Deformation and the Concept of the Continuing Loading,* Mehanika Tvjordogo Tela, No. 5, 1972, p. 16.

[3] SHANLEY, F.K., Inelastic column theory, *GAS,* Vol. 29, No. 11, 1947.

[4] RABOTNOV, Yu.N., On the equilibrium of compressed bars over proportionality limit, *Inzhenernj Slornik,* Vol. 11, 1952, pp. 123–126.

[5] PEARSON, C.F., Bifurcation criterion and plastic buckling of plates and columns, *J. Aeronaut. Sci.,* Vol. 17, No. 3, 1950.

[6] HILL, R., A general theory of uniqueness and stability of elastic-plastic solids, *J. Mechanics and Physics of Solids,* Vol. 6, No. 3, 1958, pp. 236–249.

[7] KLUSHNIKOV, V.D., The stability of the compression process of the idealized elasto-plastic rod, *Izvestja, The Academy of Sciences of the USSR, Mechanika i Mashinostroenje,* No. 6, 1964, p. 59.

[8] KLUSHNIKOV, V.D., The stability of the compression process of the idealized plate, *Mehanika*

Tvjordogo Tela, No. 4, 1966, p. 28.

[9] Rzhanitsin, A.R., The theory of creepability, *Stroj Izdat,* Moskva, 1968, pp. 213 – 236.

[10] KAMENEV, G.V., About the stability of movement at the finishing interval of time, *Prikladnaja Mechanika and Matematika,* Vol. 17, Issue 5, 1953, p. 529.

[11] RABOTNOV, Yu.N. and SHESTERIKOV, S.A., The stability of rods and plates under creepability, *Prikladnaja Mechanika i Matematika,* Vol. 21, Issue 3, 1957.

[12] KURSHIN, L.M., About the stability of rods and plates under the creepability, *The Academy of Sciences of the USSR,* Vol. 140, No. 3, 1961.

[13] RABOTNOV, Yu.N., *The Creepability of Elements and Structures,* Nauka, Moskva, 1966, pp. 200 – 205.

[14] ISHLINSKY, A.Yu., The discussion of the problem about the stability of equilibrium of the elastic solids from the point of view of elasticity, *The Ukranian Mathematical Journal,* Vol. 6, No. 2, 1954.

[15] ZHUKOV, A.M., On the question of necking in the specimen with strain, *Inzhenernj Sbornik,* Vol. 5, No. 2, 1949, pp. 103 – 107.

[16] NIKITIN, L.V. and RIZHAK, E.J., The objective laws of the demolition of rocks with internal friction and dilatancy, *Izvestja, The Academy of Sciences of the USSR, Phizika Zemly,* No. 5, 1977, p. 22.

[17] HANDELMAN, G.H., LIN, C.C. and PRAGER, W., *Quart. Appl. Math.,* Vol. 4, 1947, p. 397.

[18] DRUCKER, D.C., A more fundamental approach to plastic stress-strain relations, *Proc. of the First US Nat. Congr. of Appl. Mech., ASME,* 1951, p. 487.

[19] PRAGER, W., The stress-strain laws of the mathematical theory of plasticity – a survey of recent progress, *J. Appl. Mech.,* Vol. 15, No. 3, 1948, p. 226.

[20] KLUSHNIKOV, V.D., About the laws of plasticity for the particular class of the ways of loading, *Prikladnaja matematika i mehanika,* Vol. 21, Issue 4, 1957, p. 534.

[21] KLUSHNIKOV, V.D., New aspects in plasticity and the deformational theory, *Prikladnaja matematika i mechanika,* Vol. 23, Issue 4, 1959, p. 722.

[22] KLUSHNIKOV, V.D., The possibilities of macroexperiments and the form of governing equations, *The Reports of the Academy of Sciences of the USSR,* Vol. 264, No. 3, 1982.

[23] HILL, R., Uniqueness in general boundary-value and problems for elastic or inelastic solids, *J. of Mechanics and Physics of Solids,* Vol. 9, No. 2, 1961, pp. 114 – 130.

[24] BATDORF, S.F. and BUDIANSKY, B., A mathematical theory of plasticity based on the concept of slip, *NASA TN 1871,* 1949.

[25] VALANIS, K.S., A theory of viscoplasticity without yield surface, *Arch. Mech.,* Vol. 23, No. 4, 1977, p. 517.

[26] KADASHEVICH, Yu.J. and MICHAILOV, A.N., About the theory of plasticity, having no yield surface, *The Reports of the Academy of Sciences of the USSR,* Vol. 254, No. 3, 1980, p. 574.

Index

Fatigue and Fracture

Editor-in-Chief: Professor George C. Sih

1. S. Kocańda: *Fatigue Failure of Metals*. Revised English Edition, 1978
ISBN 90-286-0025-6
2. V.S. Kuksenko and V.P. Tamuzs: *Fracture Micromechanics of Polymer Materials*. 1981 ISBN 90-247-2557-7
3. G.C. Sih, A.J. Ishlinsky and S.T. Mileiko (eds.): *Plasticity and Failure Behavior of Solids*. Memorial Volume dedicated to the late Professor Yuriy Nickolaevich Rabotnov. 1990 ISBN 0-7923-0336-9

Kluwer Academic Publishers – Dordrecht / Boston / London

GPSR Compliance

The European Union's (EU) General Product Safety Regulation (GPSR) is a set of rules that requires consumer products to be safe and our obligations to ensure this.

If you have any concerns about our products, you can contact us on ProductSafety@springernature.com

In case Publisher is established outside the EU, the EU authorized representative is:

Springer Nature Customer Service Center GmbH
Europaplatz 3
69115 Heidelberg, Germany

Batch number: 10815974

Printed by Printforce, the Netherlands